160 Topics in Current Chemistry

W0042281

Transition Metal Coordination Chemistry

Editor: W. A. Herrmann

With contributions by
D. Astruc, J. Okuda, Ch. Zybill

With 21 Figures and 8 Tables

Springer-Verlag
Berlin Heidelberg GmbH

This series presents critical reviews of the present position and future trends in modern chemical research. It is addressed to all research and industrial chemists who wish to keep abreast of advances in their subject.

As a rule, contributions are specially commissioned. The editors and publishers will, however, always be pleased to be receive suggestions and supplementary information. Papers are accepted for "Topics in Current Chemistry" in English.

ISBN 978-3-662-14967-6   ISBN 978-3-540-47559-0 (eBook)
DOI 10.1007/978-3-540-47559-0

Library of Congress Catalog Card Number 74-644622

© Springer-Verlag Berlin Heidelberg 1992
Originally published by Springer-Verlag Berlin Heidelberg New York in 1992
Softcover reprint of the hardcover 1st edition 1992

Typesetting: Th. Müntzer, Bad Langensalza; Printing: Heenemann, Berlin;

51/3020-543210 — Printed on acid-free paper

Guest Editor

Prof. Dr. *Wolfgang A. Herrmann*
Anorganisch-chemisches Institut,
Technische Universität München,
Lichtenbergstr. 4, W-8046 Garching

Editorial Board

Preface

The present volume of *Topics in Current Chemistry* appears quite heterogeneous at first sight. Upon closer examination of the chosen topics, however, the reader will be quickly aware of the interdisciplinary approach of the three chapters dealing with modern aspects of organometallic chemistry. Low coordination numbers, steric bulkiness, and π-coordination are important motives and guidelines in the design of new and unseen molecular compounds. For example, stabilization of low-coordinated species is normally achieved by sterically demanding ligands. In this context, J. Okuda demonstrates interligand interactions and conformational analysis relevant to sterically demanding cyclopentadienyl ligands attached to transition metals. Peripheral steric bulk has an enormous influence upon the reactivity of the metal centers in these complexes.

Substitution effects of π-aromatic ligands are also highlighted when the formation of CC-bonds — one major aspect of applied organometallic chemistry — is to be directed by means of electrochemical effects. In this context, D. Astruc has established spectacular tentacle molecules by multiple deprotonation/alkylation sequences, a beautiful example of "The Use of π-Organoiron Sandwiches in Aromatic Chemistry". While the organic chemistry of transition metals has exploited silyl-containing ligands and substituents mostly for steric and other stabilization effects in the periphery of molecules, metal-to-silicon multiple bonds have long and unsuccessfully been looked for by many workers. Only Ch. Zybill has presented a straightforward synthesis of silylene-metal complexes, e.g. $M = Si$ double bonds, the heavier congeners of Fischer's metal carbenes of 1964. Zybill's report on "The Coordination Chemistry of Low Valent Silicon" highlights the specific chemistry of divalent silicon (silylenes) at transition metal sites for the first time.

It can be readily seen that all three areas summarized in this volume will very soon also play a major role in applied organometallic chemistry.

Wolfgang A. Herrmann
Technische Universität München

Table of Contents

The Coordination Chemistry of Low Valent Silicon

Christian Zybill

Anorganisch-chemisches Institut, Technische Universität München, Lichtenbergstr. 4, W-8046 Garching, FRG

Table of Contents

Topics in Current Chemistry, Vol. 160
© Springer-Verlag Berlin Heidelberg 1991

Christian Zybill

The organometallic chemistry of silicon is experiencing a remarkable period of renaissance. Particularly coordination compounds with low valent silicon ligands are becoming more and more the focus of interest. This refers not only to silicon clusters, polymers or amorphous silicon, but also to highly reactive intermediates such as silylenes (silanediyls) or disilenes. Reactive species can be fixated in the coordination sphere of a transition metal and a further complexation of the silicon by a donor leads to surprisingly stable compounds. Such silicon congeners of carbene, olefin and similar complexes have long been sought for and in many cases have only been known as reactive intermediates of silylene transfer reactions, of metal catalyzed Si—Si bond formation reactions or e.g. the hydrosilation process. The introduced stable donor adducts now allow a facile exploration of the chemical potential of these compounds. A lot of implementations are given for applications, and spin off results might reach as far as to microelectronics and semiconductor technology. A report of the current state of this relatively new research area is given and covers the literature until 1990.

1 Introduction

Rarely has any research area so continuously been in the focus of scientific interest as organosilicon chemistry. A rapidly increasing number of publications, review articles, and scientific conferences reflects this development. The initiating factor was the spectacular discovery of stable molecules with SiSi double bonds, disilaethenes (disilenes) by R. West and S. Masamune in 1981 and 1982 [1–7].

This fundamental discovery dramatically affected the whole chemistry of main-group elements. Subsequently, a series of new compounds with silicon element multiple bonds has been introduced. Within only a few years, stable silenes (silaethenes with a $Si=C$ double bond) [8–11], silaimines $Si=N$ [12–14], and silaphosphenes $Si=P$ [15] were synthesized. As a pacemaker, silicon chemistry has exerted a strong influence on further areas of main-group chemistry; a variety of stable molecules with $Ge=Ge$ [16], $P=P$ [17], $As=As$ [18], $P=C$ and $P\equiv C$ [19–22] bonds were subsequently isolated, and systems with cumulated double bonds $P=C=P$ [23–25] are also known today.

In the course of this development, knowledge about low valent (in the sense of formal low oxidation states) reactive intermediates has significantly increased [26–30]. On the basis of numerous direct observations of silylenes (silanediyles), e.g., by matrix isolation techniques, the physical data and reactivities of these intermediates are now precisely known [31]. The number of kinetic studies and theoretical articles on reactive intermediates of silicon is still continuously growing [32–34].

Very recently, the coordination chemistry of low valent silicon ligands has been established as an independent, rapidly expanding research area. With the discovery of stable coordination compounds of silylenes [35–38], a major breakthrough was achieved. Within a short time a variety of stable complexes with a surprising diversity of structural elements was realized. Besides neutral coordination compounds (**A, B**) [35, 36, 38], and cationic compounds (**C**) [37], also cyclic bissilylene complexes (**D**) [39, 40] exist. A common feature of the above-mentioned compounds is the coordination of an additional stabilizing base (solvent) to the silicon. However, "base-free" silylene complexes (**A**) are also accessible as reactive intermediates at low temperatures.

The coordination sphere of transition-metal complexes can furthermore be utilized for the fixation of silicon ligands in their lowest oxidation states. Even examples of compounds containing a formally zerovalent silicon (**E**) are now known [41].

Also π-complexes of silaethenes (**F**) and disilaethenes (**G**) have been realized recently, which in a wider sense can be classified as coordination compounds of low valent silicon ligands. Some examples are briefly discussed in the two final sections of this article [42–48].

"Donor free" silylene complexes are reactive intermediates in a variety of chemical reactions. In many cases, evidence for the coordinated silylenes involved has been obtained indirectly by means of trapping experiments [49–60].

With the stable donor adducts of silylene complexes, valuable model compounds are now available for reactive intermediates which otherwise cannot be observed directly. For example, a side reaction occurring in the hydrosilation process [61–63], is the dehydrogenative coupling of silanes to disilanes. This reaction could be explained in terms of a silylene transfer reaction with a coordinated silylene as the key intermediate.

A variety of further reactions are known in which silylenes are transferred to a substrate in the presence of a transition-metal catalyst. In most cases, silylene complexes can now be identified as reactive intermediates. For a detailed discussion refer to Sects. 2.5.3 and 2.5.4.

Investigations of silicon-metal systems are of fundamental interest, since stable coordination compounds with low valent silicon are still rare [64], and furthermore, silicon transition-metal complexes have a high potential for technical applications. For instance, coordination compounds of Ti, Zr, and Hf are effective catalysts for the polymerization of silanes to oligomeric chain-silanes. The mechanism of this polymerization reaction has not yet been fully elucidated, but silylene complexes as intermediates have been the subject of discussion. Polysilanes find wide use in important applications, e.g., as preceramics [65–67] or as photoresists [68–83].

The growing interest in volatile silyl-metal complexes for chemical vapor deposition reactions should also be mentioned. This technique is extremely useful for the preparation of silicide films in microelectronic devices. Further examples of applications of silicon-metal compounds are given in the appropriate sections.

2 Monomeric Base-Stabilized Silylene Complexes

"Does a Silylene-Complex exist?" This rhetorical question is the title of a theoretical paper published in 1983 [84]. As a result of an ab-initio calculation, the authors came to the conclusion that a moderately positive answer can be given. However, silylene complexes are thermodynamically less stable than carbene complexes (the MSi bond energy for the hypothetical complex $(OC)_5Cr=Si(OH)H$ is 29.6 kcal/mol, the bond energy of the MC bond in $(OC)_5Cr=C(OH)H$ is 44.4 kcal/mol) [85], and therefore silylene complexes should be difficult to isolate.

This argument is a typical example of the state of knowledge of silylene complexes until 1987, when stable compounds were only known in form of dimeric complexes. Accordingly, the number of publications related to this subject is rather high. Very recently, cyclic bis-silyleneplatinum compounds again became the focus of scientific interest, since in some Pt-four-membered ring complexes surprisingly short cross ring SiSi distances were observed. These features have been interpreted in terms of a weak SiSi interaction (Sect. 3).

Much effort has been invested into the search for monomeric compounds [86–90]. One of the few reliable indications for a base-stabilized monomeric complex involved the use of spectroscopic methods. The complexes 1 and 2 could be isolated and characterized below $-20\,°C$ with spectroscopic methods [91].

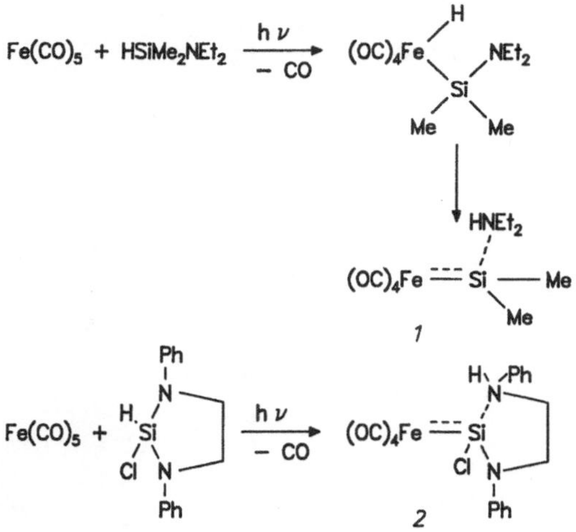

$$(1)$$

In 1987 a major breakthrough was achieved when two research groups independently succeeded in the synthesis of monomeric silylene complexes in the form of stable base adducts [35–38].

A simple synthetic approach proved to be as most effective: Reaction of $Na_2Fe(CO)_4$ with $(t\text{-BuO})_2SiCl_2$ provides the desired product 4 as the solvent-complex bis(t-butoxysilanediyl)tetracarbonyliron(0) × HMPA. This compound has been synthesized in our laboratories and is the first stable coordination compound of a silylene (Fig. 1a, b, c) [35]. The x-ray structure analysis of 4 shows the silylene ligand occupying an apical coordination site of the trigonal bipyramid at the iron center. The silicon atom is further complexed via the oxygen atom by an additional HMPA molecule. Together with the two t-BuO substituents, a distorted tetrahedral coordination geometry is formed at the silicon, which is rather similar to observed geometries of base adducts of silaethenes and silaimines. The bond distance (HMPA)-oxygen silicon (1.730(3) Å, Si−OtBu 1.610(3)/1.636(4) Å) is significantly enlarged and indicates a loose coordination of the solvent. With 2.289(2) Å also the Fe−Si bond is shortened. This coordination geometry justifies a description of 4 as a base-stabilized silylene complex.

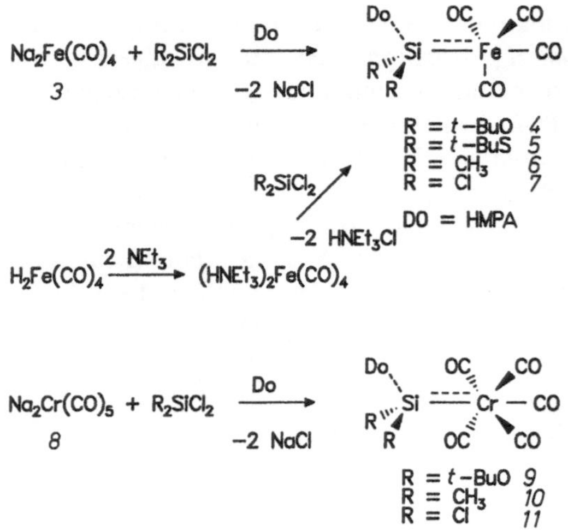
Do
$Na_2Fe(CO)_4$ + R_2SiCl_2
3
−2 NaCl
Do OC CO
Si Fe CO
R R CO
R = t-BuO 4
R = t-BuS 5
R = CH_3 6
R = Cl 7
DO = HMPA
R_2SiCl_2
−2 $HNEt_3Cl$
$H_2Fe(CO)_4$ 2 NEt_3 $(HNEt_3)_2Fe(CO)_4$
$Na_2Cr(CO)_5$ + R_2SiCl_2
8
Do
−2 NaCl
Do OC CO
Si Cr CO
R R OC CO
R = t-BuO 9
R = CH_3 10
R = Cl 11

(2)

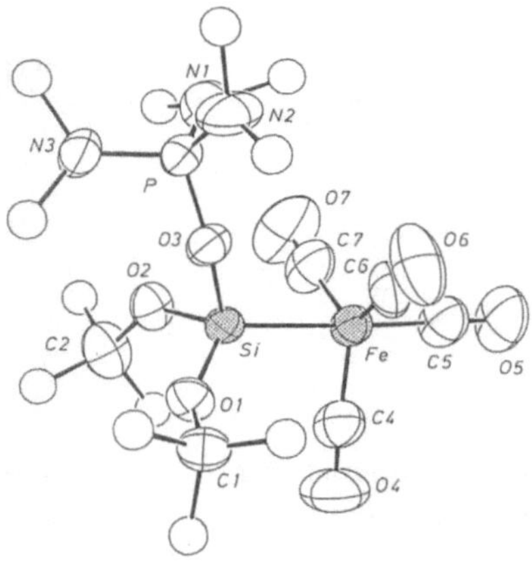
N1
N2
N3
P
O3
O2
O7
C7
C6
O6
C2
Si
Fe
C5
O5
O1
C4
C1
O4

a

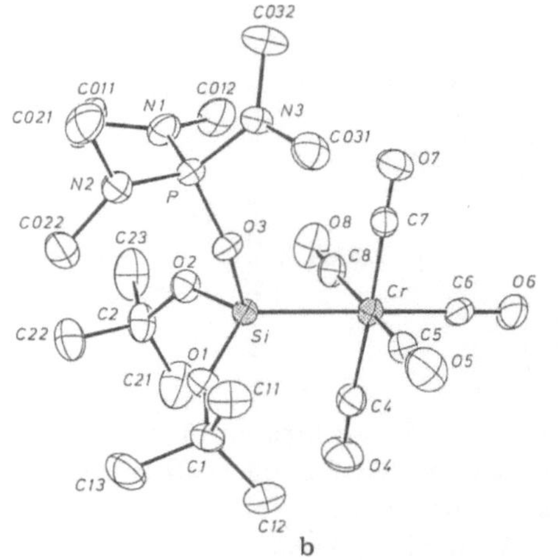
C032
C011
C012
N1
N3
C021
C031
O7
N2
P
C022
O3
O8
C7
C23
C8
O2
C2
Cr
C6
O6
C22
Si
C5
O1
O5
C21
C11
C4
C1
O4
C13
C12

b

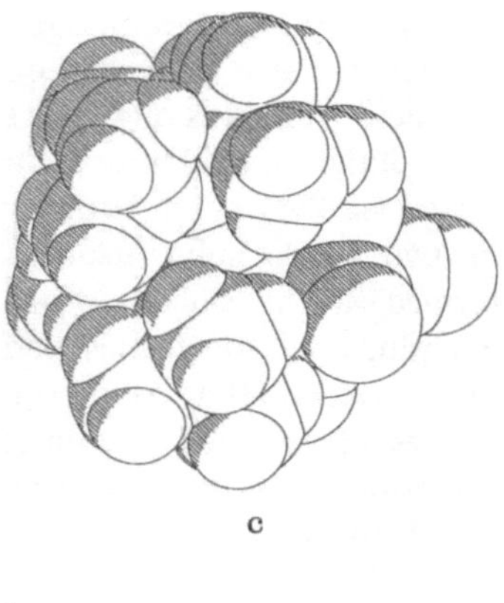

c

Fig. 1a. Structure of the complex bis(*t*-butoxysilanediyl)tetracarbonyliron(0) × HMPA **4**. **b.** Molecular structure of the chromium complex bis(*t*-butoxysilanediyl)pentacarbonylchromium(0) × HMPA **9**. **c.** Van der Waals surface of **9**

The synthetic approach to silylene complexes (Eq. (2)) is versatile and allows a high variability of both the metals and the substituents at the silicon. A whole series of compounds with bulky substituents like 1-adamantyloxy, 2-adamantyloxy, neopentyloxy, triphenylmethoxy or t-butylthio could be prepared (Table 1). Compounds with sulfur at silicon are particularly interesting; however, their synthesis proved to be very difficult.

With regard to the stabilizing effect of the α-substituent at the silicon, the following gradation can be inferred from results of x-ray structures: O > S > C > Cl. This sequence correlates with known Si-X bond energies.

Silylene complexes are not only stable with donor substituents but also with simple alkyl residues at silicon. These alkyl complexes still have a sufficient thermodynamic stability, but otherwise are reactive enough to allow a rich and diverse chemistry. Particularly the chlorocompounds 7 and 11 are valuable starting materials for further functionalization reactions; the details of these reactions will be discussed in the forthcoming sections. The data for the known compounds are summarized in Table 1.

The coordinated silylenes in both the iron and the chromium compounds can be photolytically activated: Photolysis of the complexes in the presence of triphenylphosphine gives the *trans*-silylene-phosphine complex, which in a second step is transformed into the *trans*-bisphosphine compound by excess phosphine. If the silylenes are not trapped, polysilanes are isolated in almost quantitative

Table 1. ^{29}Si-NMR data of selected compounds

Silylene complexes Compound		δ/ppm	$^2J(^{31}P^{29}Si)$/Hz	Ref.
$(OC)_4FeSi(HMPA)(t\text{-}BuO)_2$	**4**	7.1	26.4	[35]
$(OC)_4FeSi(HMPA)(t\text{-}BuS)_2$	**5**	74.7	25.3	[143]
$(OC)_4FeSi(HMPA)Me_2$	**6**	92.3	17.5	[38]
$(OC)_4FeSi(HMPA)Cl_2$	**7**	49.7	31.2	[38]
$(OC)_5CrSi(HMPA)(t\text{-}BuO)_2$	**9**	12.7	37.2	[36]
$(OC)_5CrSi(HMPA)Me_2$	**10**	101.4	31.3	[38]
$(OC)_5CrSi(HMPA)Cl_2$	**11**	55.0	41.4	[38]
$Cp^*Ru(PMe_3)_2SiPh_2^+\ BPh_4^-$	**12**	95.8		[37]
$(OC)_5CrSi(HMPA)(1\text{-}AdaO)_2$	**16**	11.9	30.1	[143]
$(OC)_5CrSi(HMPA)(2\text{-}AdaO)_2$	**17**	11.7	30.2	[143]
$(OC)_5CrSi(HMPA)(neopentylO)_2$	**18**	12.5	31.1	[143]
$(OC)_5CrSi(HMPA)(tritylO)_2$	**19**	10.9	32.0	[143]
$(OC)_5CrSi(HMPA)(t\text{-}BuS)_2$	**20**	88.2	31.0	[143]
$(OC)_4FeSi(t\text{-}BuS)_2$	**21**	83.2	–	[143]
$(OC)_4FeSi(HMPA)_2Fe(CO)_4$	**22**	24.1	22.0	[41]

Cyclic compounds	δ/ppm		Ref.
15	121.1/101.9		[40]
$[(OC)_4CrSiMe_2]_2$ **55**	159.0		[143]

Christian Zybill

Table 1 (continued)

Miscellaneous compounds	δ/ppm	δ/ppm	Ref.
FpSiMePhSiMe₃	−11.6 (SiMe₃)	12.7 (FeSi)	[139]
FpSiMe₂SiMe₂Ph	−15.0 (SiMe₂Ph)	12.7 (FeSi)	[139]
FpSiMePhSiMe₂Ph	−15.4 (SiMe₂Ph)	12.3 (FeSi)	[139]
FpSiMe₂SiMePh₂	−18.2 (SiMePh₂)	17.0 (FeSi)	[139]
FpSiMePhSiMePh₂	−16.0 (SiMePh₂)	12.2 (FeSi)	[139]
FpSiMe₂SiPh₃ **37**	−15.4 (SiPh₃)	16.8 (FeSi)	[139]
dcpePt(Cl)Si(TMS₂)Ph	−34.46 (PtSi)	−9.70 /TMS)	[115]
FpSi(C₄Ph₄)Me **34**	47.59		[136]
FpSi(C₁₀H₄Ph₄)Me **35**	138.98		[136]
FpSi(C₆Ph₄(COOMe)₂ **36**	166.56		[136]
Cp₂Ti(μ−SiH₂Ph)(μ−H)TiCp₂ **63**	87.2		[117]
C₆Me₆(CO)₂Cr(H)SiHPh₂ **30**	21.2		[130]
Cp₂Zr(SiHMes₂)Me	−12.36		[116]
CpCp*Zr(Cl)SiH₂Ph	−14.33		[116]
CpCp*Zr(Cl)Si(SiMe₃)₃ **28**			[146]
CpCp*Hf(Cl)Si(SiMe₃)₃ **29**			[146]

Silene complexes	δ/ppm	Ref.
Cp*Ru(PiPr₃)H(CH₂SiMe₂)	6.14	[44]

Disilene complexes	δ/ppm	Ref.
dppePt(SiiPr₂SiiPr₂)	19.60	[47]
Cp₂W(SiMe₂SiMe₂)	−48.1	[48]

yield. In all cases the liberated HMPA can be recovered. This reactivity clearly documents the potential of the introduced complexes to react as silylene sources [38].

As well as the neutral compounds, also cationic complexes are available from chlorosilyl compounds by chloride abstraction. One example **12** has been characterized by x-ray structure analysis (Fig. 2).

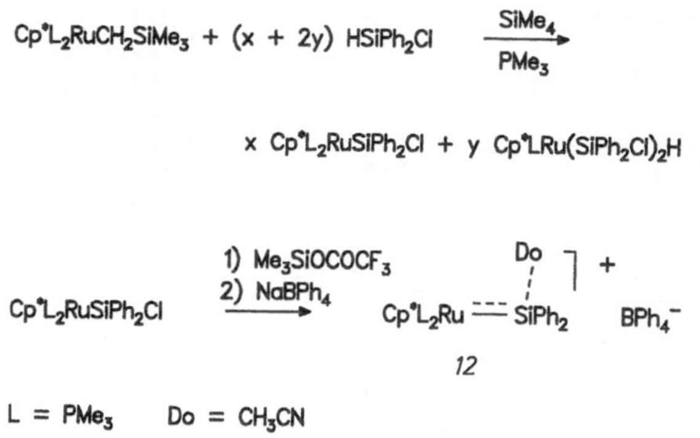

$$Cp^*L_2RuCH_2SiMe_3 + (x + 2y)\ HSiPh_2Cl \xrightarrow[PMe_3]{SiMe_4}$$

$$x\ Cp^*L_2RuSiPh_2Cl + y\ Cp^*LRu(SiPh_2Cl)_2H$$

(3)

$$Cp^*L_2RuSiPh_2Cl \xrightarrow[\text{2) NaBPh}_4]{\text{1) Me}_3SiOCOCF_3} \left[Cp^*L_2Ru \overset{Do}{=\!=\!=} SiPh_2 \right]^+ BPh_4^-$$

12

L = PMe₃   Do = CH₃CN

8

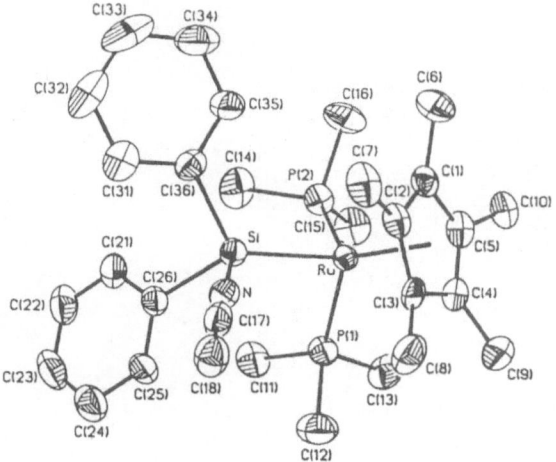

Fig. 2. ORTEP view of the cation in bis(trimethylphosphine)(diphenylsilanediyl)(pentame-thylcyclopentadienyl)ruthenium tetraphenylborate × acetonitrile **12** [37]

In this case also the silicon is four-coordinated as a result of the supplementary complexation by acetonitrile. With a distance of 2.328(2) Å the RuSi bond can be considered as short. It is interesting to note that **12** is the only case in which a reversible coordination of the donor molecule at the silicon has been found [92]. From these results a high reactivity of **12** should be inferred; however, no details are known so far [37].

Very recently, even examples of donor-stabilized bissilylene complexes have been introduced. For instance, the two cyclic systems **13** and **14** are prepared by an aminolysis reaction of the respective bis(chloro)silyl complexes. The x-ray structure analysis of **13** has been performed [39].

$$Cp(OC)Fe \underset{H_3C\ CH_3}{\overset{H_3C\ CH_3}{\langle\!\langle\ \rangle\!\rangle}} NMe_2 \qquad Cp(OC)_2Fe \underset{H_3C\ CH_3}{\overset{H_3C\ CH_3}{\langle\!\langle\ \rangle\!\rangle}} N^tBu \qquad (3a)$$

13   *14*

Access to the oxo-substituted complex **15** has been achieved by means of elegant photochemical methods. The structure of **15** shows a close analogy (Fe−Si(1) 2.222(3), Fe−Si(2) 2.207(3) Å) to the comparable N-compound. The methoxy bridge is unsymmetrical (Si(1)−O(2) 1.793(9), 1.799(8) and Si(2)−O(3) 1.632(9) Å) (Fig. 3). The relatively weak coordination of the OMe residue can be explained by a strong donor effect of the Cp* ring. No reactions of this compound have been reported.

Christian Zybill

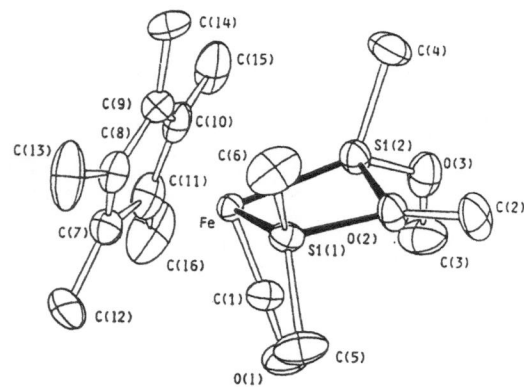

Fig. 3. X-ray structure of the cyclic bissilylene complex **15** CpFe(CO)−SiMe$_2$−OMe−SiMeOMe [40]

$$\text{Cp*Fe(CO)}_2\text{−SiMe}_2\text{−SiMe(OMe)}_2 \xrightarrow[\text{−CO}]{h\nu}$$

(4)

15

Considering the interesting bissilylene complexes, the question arises how many silylenes can be coordinated to a transition metal. In the series of stannylene complexes, a maximum coordination number of at least 3 could be established.

2.1 Methods for the Synthesis of Silylene Complexes

The first silyl-metal complex was synthesized in 1956 with the compound CpFe(CO)$_2$SiMe$_3$ [93, 94]. Since then, methods for the formation of M-Si bonds have been investigated systematically and developed to an effective set of tools. Numerous articles covering this subject give evidence for the continuous development of this area [64, 95].

Basically the same methods known from the synthesis of classical metal-silyl complexes can also be applied to the preparation of low valent Si compounds. The procedures given here are summarized with the focus on silylene complexes: These are a) reactions of appropriate metal anions with halosilanes, which are the most important methods for the formation of M-Si bonds. Alternatively, silyl

10

anions can be treated with metal halides. b) The large area of oxidative addition reactions of, e.g., silanes to coordinatively unsaturated metal complexes which open up new elegant routes for the formation of metal silicon bonds. These reactions can be accompanied by the successive step of a reductive elimination reaction, so that the primary reaction products of silylmetal hydrides are not isolated. c) Finally, photochemically or thermally generated silylenes can be trapped by 16 e transition-metal fragments. This has proved to be particularly effective for the direct generation of silylenes in the coordination sphere of a metal.

2.1.1 Reactions of Metal Anions with Halosilanes

Carbonylate anions are the most suitable starting material for the synthesis of silylmetal compounds. A prerequisite for the preparation of compounds with a formal $M = Si$ double bond is the use of metallate dianions like $Na_2Fe(CO)_4$ (Collman's reagent) together with the respective dichlorosilanes [96].

Investigations of carbonylate polyanions have been intensely pursued by Ellis and co-workers [97–107].

Most of the compounds have been found to react as supernucleophiles.

The chlorosilanes are available from classical synthetic routes [108]. In most cases the products can be obtained from $SiCl_4$ and the lithium salts of the introduced substituents with sufficient selectivity and in high yield. This is particularly true for the alcoholates and the thiolates. The exchange problems R vs. Cl which arise with the t-butylthiochlorosilanes will not be discussed in detail here. These problems are overcome by an appropriate choice of the reaction conditions.

It is known that $Na_2Fe(CO)_4$ can be silylated twice to form cis-$[(H_3C)_3Si]_2Fe(CO)_4$ [109]. Also the reaction of $Na_2Fe(CO)_4$ with 1.1-dichlorosilanes has been described and leads exclusively to the dimeric compounds [110, 111]. In polar solvents the formation of dimers can be suppressed and monomeric base-stabilized compounds are obtained. A very elegant procedure is the in-situ generation of the carbonylate anions in solution by deprotonation of $H_2Fe(CO)_4$.

$$Na_2Fe(CO)_4 + R_2SiCl_2 \xrightarrow[-\ NaCl]{k_1} \left[\begin{array}{c} (OC)_4Fe \longrightarrow SiR_2 \\ \vdots \qquad | \\ Na \; \text{--} \; Cl \end{array} \right]$$

$$\Big\downarrow \begin{array}{c} Do \\ -\ NaCl \end{array} \; k_2$$

$$(OC)_4Fe \overset{Do}{\underset{}{=\!=\!=}} SiR_2 \qquad\qquad (5)$$

R = t–BuO, t–BuS, MesO,
1–AdaO, 2–AdaO, CH_3, Cl   Do = THF, HMPA

The reaction of the carbonyl metallates with chlorosilanes (Eq. (5)) can be considered to be a nucleophilic displacement at the silicon followed by NaCl elimination. Several arguments support this view: First of all, strong dipolar aprotic solvents are essential, and furthermore, no indication for the possible involvement of silylenes or for any silyl radicals (negative trapping experiments) could be obtained. However, it should be noted that k_2 must be much larger than k_1 since no intermediate could be detected.

The reaction (Eq. (5)) in THF yields labile THF adducts which are converted into the more stable HMPA adducts by addition of HMPA. The various equilibria existing between $Na_2Fe(CO)_4$ and several donor solvents are described in a detailed paper by Collman: in HMPA, the solvent-separated supernucleophilic ion pair $[Na^+ \times HMPA \times Fe(CO)_4^{2-}]^-$ is the kinetically dominant species, with no kinetic contribution from free $[Fe(CO)_4]^{2-}$. In THF, $Na_2Fe(CO)_4$ is much less dissociated, with tight-ion paired $[NaFe(CO)_4]^-$ as the kinetically important species [96].

A complex $Na[Na \times HMPA \times Fe(CO)_4]$ could be isolated by us from HMPA in crystalline form.

A further important feature of HMPA is its stabilizing effect on the Redox potential of $[Fe(CO)_4]^{2-}$ by ion solvation. In less polar solvents, electron-transfer reactions take place and $[Fe(CO)_4]^{2-}$ is oxidized to $[HFe_3(CO)_{11}]^-$. This redox reaction is suppressed in HMPA.

As can be seen from Table 1, a whole set of compounds is available from the reaction of Eq. (5), which shows the high variability of both the substituents at silicon and the introduced metal. This method is only limited by the steric demand of the substituents. If the residues at the silicon are too bulky, no reaction according to Eq. (5) occurs. The method also allows the use of dinuclear carbonylate anions (Eq. (6)) [112]:

$$R_2SiCl_2 + Na_2M_2(CO)_{10} \xrightarrow[- [M(CO)_5]]{Do} R_2Si \overset{Do}{\underset{OC}{\overset{OC}{=}}} \overset{OC}{\underset{CO}{}} M-CO \tag{6}$$

$$R = t\text{-BuO}$$
$$M = Cr, Mo, W$$

An example of particular interest is the two-fold introduction of $M(CO)_n$ moieties at silicon to give HMPA adducts of organometallic analogues of silaallene. It has been shown that this reaction proceeds through the dichlorosilylene complex as intermediate. Both the iron 22 and ruthenium 23 compound and also the bimetallic complex 24 are accessible.

$$2\ Na_2M(CO)_n + SiCl_4 \xrightarrow[- 4\ NaCl]{2\ Do} (OC)_nM_1 \overset{Do\ Do}{=\!=\!=} Si \overset{}{=\!=\!=} M_2(CO)_n$$

$$\begin{aligned}
M_1 &= Fe, M_2 = Fe \quad 22 \\
M_1 &= Ru, M_2 = Ru \quad 23 \\
M_1 &= Fe, M_2 = Ru \quad 24
\end{aligned} \tag{7}$$

2.1.2 Reactions of Silyl Anions with Metal Halides

An interesting variant of metal-silicon bond formation is the combination of metal halides with silyl anions. Since silyl dianions are not available, only one metal-silicon bond can be formed directly. The silylene complexes are then accessible by subsequent reaction steps [113]. An example of this approach is given by the reaction of cis-bistriethylphosphaneplatinumdichloride 25 with diphenylsilylli-thium, which yields, however, only dimeric platinadisilacyclosilanes 26a–c [114].

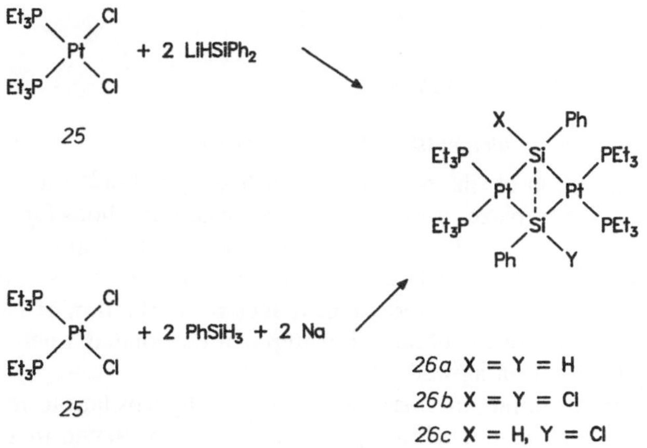

$$\text{(8)}$$

26a X = Y = H

26b X = Y = Cl

26c X = H, Y = Cl

The reaction of (bistrimethylsilyl)phenylsilyllithium with bis(dicyclohexylphos-phino)ethaneplatinumdichloride also does not lead to monomeric silylene complexes but only to the silylplatinum compound 27b [115].

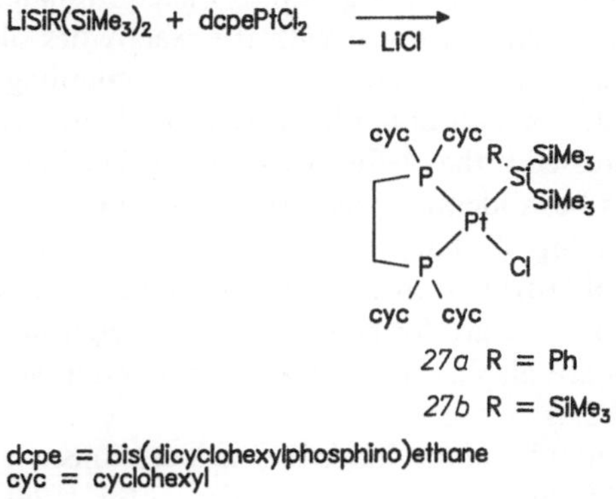

$$\text{(9)}$$

27a R = Ph

27b R = SiMe$_3$

dcpe = bis(dicyclohexylphosphino)ethane
cyc = cyclohexyl

Finally, the reaction of silyl anions with dichlorides of Zr and Hf, which provides the silyl complexes 28 and 29, should be mentioned [116].

Christian Zybill

$$CpCp^*MCl_2 + LiSi(SiMe_3)_3 \longrightarrow$$

$$CpCp^*M\begin{matrix} Cl \\ \diagdown \\ Si(SiMe_3)_3 \end{matrix} \qquad (10)$$

$$M = Zr \quad 28$$
$$M = Hf \quad 29$$

These and similar complexes of Ti and Zr are effective catalysts for the formation of polysilanes from primary silanes [117–120].

2.1.3 Oxidative Addition Reactions

2.1.3.1 Reactions of Silanes with Transition Metals

The oxidative addition of silanes (with silicon-hydrogen bonds) to coordinatively unsaturated metal complexes is one of the most elegant methods for the formation of metal-silicon bonds. Under this heading normally reactions are considered which yield stable silyl metal hydrides. However, in some cases the oxidative addition is accompanied by a subsequent reductive elimination of, e.g., hydrogen, and only the products of the elimination step can be isolated. Such reactions are considered in this section as well.

An important example in this context is the hydrosilation reaction [121]. Hydrosilation is the formal addition of a silane to an alkene in presence of a hydrosilation catalyst. This reaction has numerous applications; e.g., with a suitable catalyst, an enantioselective hydrosilation is possible [122] and also hydrosilation and double hydrosilation of alkines [123] are known.

A great number of articles related to the mechanism of this reaction has been published. It can be considered as certain that the silanes react with the platinum center by an oxidative addition to the metal with formation of a silylplatinum hydride and subsequent transfer of the silyl group to the coordinated alkene.

Recent investigations have been concerned with the reactivities observed with secondary silanes R_2SiH_2. In these cases, a dehydrogenative coupling of silanes to disilanes is observed as a side reaction of the hydrosilation. However, the hydrosilation can be totally suppressed if the olefins are omitted. The key intermediate in the coupling reaction has been identified as a silylene complex (sect. 2.5.4).

The dehydrogenative coupling of silanes does not stop at the stage of disilanes in the coordination sphere of "early" transition metals like Zr and Hf, but chain polymers of low molecular weight are formed. As reactive intermediates in this reaction, silylene complexes are also assumed. However, alternative mechanisms have been discussed (sect. 2.5.4).

Particularly interesting examples of mechanistic considerations can be found in the work of Graham [124–126], Schubert [127, 128] and Crabtree [129]. Cophotolysis of hexamethylbenzenechrom- (or tungsten) tricarbonyl or $Cp^*Mn(CO)_3$ with diphenylsilane yields, after cleavage of CO, a "side-on coordinated" silane with a M(H)Si 3c2e bond **30** [130, 131].

14

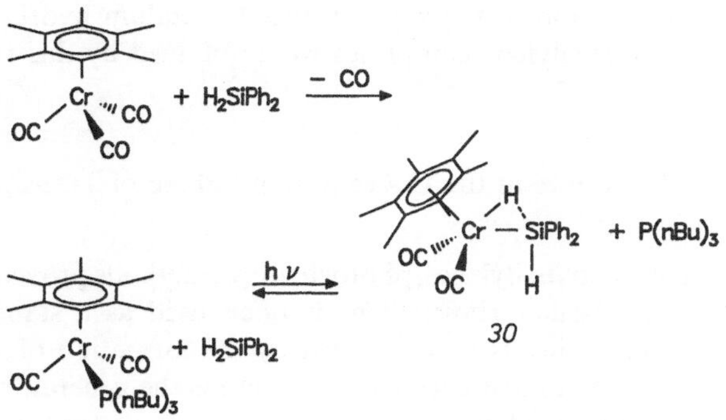

$$(11)$$

The results of x-ray structure analysis and neutron diffraction, as well as spectroscopic experiments ($J(HSi) = 70.8$ Hz for **30**), can be interpreted in the sense mentioned above. The observed reactivity of **30** is also consistent with this view, the coordinated silanes can be displaced smoothly by phosphines, according to first-order reaction kinetics.

However, recently, a theoretical paper has been published, which provides interesting arguments for a conventional silylmetal hydride rather than for a 3c2e M(H)Si bond [132, 133]. For the great implications of these compounds for Si — H bond activation reactions consult, e.g., on the work of Crabtree [129].

2.1.3.2 Insertion of Silylenes into Metal-Hydrogen Bonds

Relatively few examples are known which utilize an oxidative addition reaction of metal hydrides to necessarily low valent silicon compounds. Seyfert's hexamethylsilirane (**31**) could be used as a source of dimethylsilylene to perform an

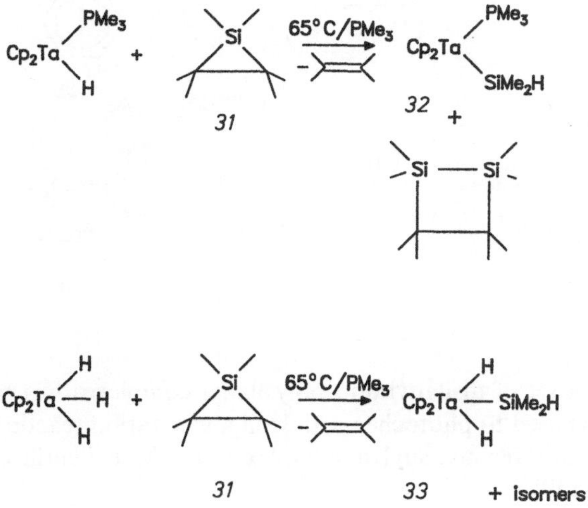

$$(12)$$

15

Christian Zybill

interesting insertion reaction of the silylene into a tantalum hydride bond [134, 135]. However, no stable silylene complexes were obtained by this route.

2.1.4 Generation of Silylenes in the Coordination Sphere of Transition Metals

For the generation of reactive silylenes, photochemical methods proved particularly effective. For instance, 1-silanorbornadiene is often used as a standard silylene source. The silylene extrusion is accompanied by the formation of stable arenes (rearomatization energy). A most effective procedure is the generation of silylenes directly in the coordination sphere of the metal. An example for this strategy has been given by R. West. The precursor for the photochemical reaction is prepared by a classical route by means of the reaction of the $[CpFe(CO)_2]^-$ anion with a chlorosilane. **34** is converted to the corresponding silanorbornadiene complexes by reaction with benzines or alkines, respectively. Upon photolysis, the sila-norbornadiene complexes provide reactive ferra-substituted silylenes, which can be trapped by a variety of trapping reagents [136].

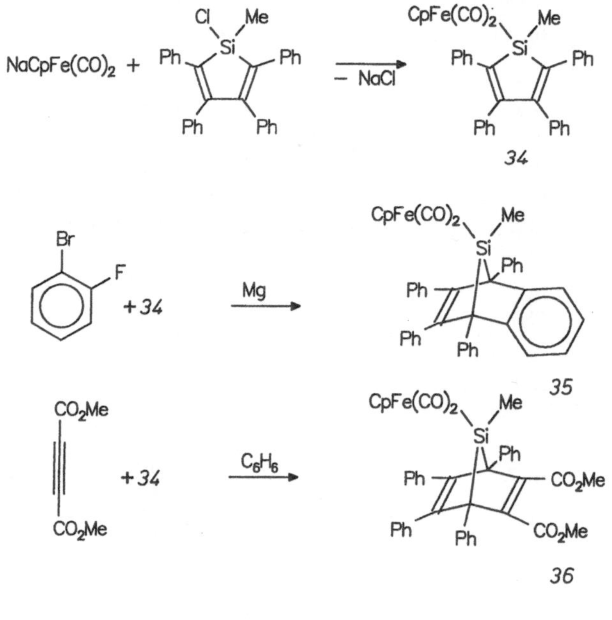

(13)

In a similar way, a set of disilanyl and polysilanyl complexes has recently been synthesized and exposed to photochemical deoligomerization reactions [137]. The photolytically obtained reactive silylene complexes have been identified by trapping experiments [138, 139].

16

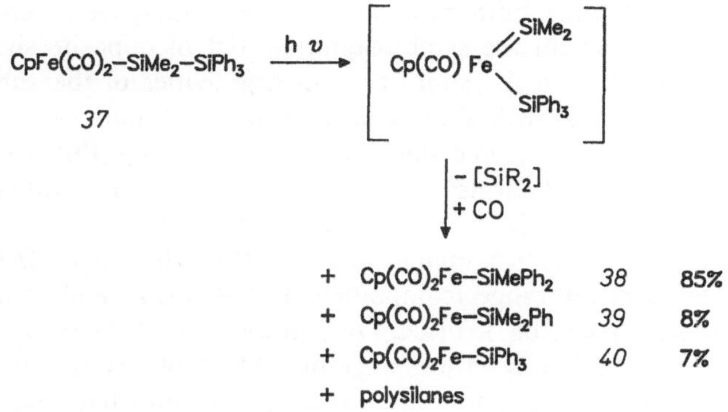

$$CpFe(CO)_2-SiMe_2-SiPh_3 \xrightarrow{h\,\upsilon} \left[Cp(CO)\,Fe \diagdown\!\!\!\overset{\diagup SiMe_2}{\diagdown SiPh_3} \right]$$

37

$$\Big\downarrow \begin{array}{l} -[SiR_2] \\ +CO \end{array} \qquad\qquad (14)$$

+ Cp(CO)$_2$Fe–SiMePh$_2$ 38   85%

+ Cp(CO)$_2$Fe–SiMe$_2$Ph   39   8%

+ Cp(CO)$_2$Fe–SiPh$_3$ 40   7%

+ polysilanes

$$Cp(CO)_2Fe-SiMe_2-SiMe_3 \longrightarrow Cp(CO)_2Fe-SiMe_3 \qquad\qquad (15)$$

41   + polymer   42

2.2 Spectroscopic Data of Silylene Complexes

2.2.1 NMR Spectroscopy

^{29}Si-NMR spectroscopy is a most effective method for the structure assignment of silicon compounds in solution. The classical pulse methods are still important, they should, however, carefully be adapted to each particular case. Considerable relaxation problems for metal-substituted silicon compounds make it necessary to use very low pulse angles (10°); this measure drastically shortens the pulse delay intervals and allows a gain of higher repetition rates. But severe problems with NOE effects can only be overcome by the use of modern pulse sequence programs as INEPT, DEPT, etc. In those cases in which a J(SiH) coupling of appropriate magnitude is available for a polarization transfer from hydrogen to silicon (other examples are not known), an essential enhancement of the signal/noise ratio can be achieved. The significance of modern pulse techniques for ^{29}Si-NMR spectroscopy has been shown in numerous examples [140]. Recently, also 2D-correlated INEPT ^{29}Si-NMR spectroscopy has been introduced [141].

In Table 1 the ^{29}Si-NMR data of stable silylene complexes and related compounds are summarized. The ^{29}Si-NMR data cover a wide chemical shift range which is clearly distinguished from the shift values of silyl metal compounds. The influence of the substituents on the silicon is evident and shows the high diagnostic value of ^{29}Si-NMR spectroscopy. In the series of HMPA adducts, a characteristic trend can be found which is quite similar to that observed with regular tetravalent compounds of silicon. These trends can be explained in terms of paramagnetic and diamagnetic shift influences on the silicon nucleus [142].

17

According to calculations, both paramagnetic and diamagnetic contributions to the ^{29}Si-NMR shift are of the same magnitude but of opposite sign. Thus, no linear correlation between the sum of electronegativities of the substituents at silicon and the observed NMR shift exists, and the NMR data cannot directly be interpreted in terms of electronic shielding and deshielding. But with a detailed knowledge of all involved effects, ^{29}Si-NMR spectroscopy is a valuable method for structure assignment (Fig. 4).

From the invariance of the coupling constant ^{2}J(SiP) between HMPA-P and Si over a wide temperature range (compounds 4–7, 9–11) a rigid coordination of the HMPA to silicon can be deduced. Only in the case of the methyl complex 6 above 25 °C is the beginning of exchange of HMPA observed. However, a fast exchange of the coordinated acetonitrile at room temperature has been found for 12.

For 6, the activation energy for rotation about the MSi bond has been measured as $\Delta G = 40.3(\pm 5)$ kJ/mol [143]. According to MO calculations, a genuine Cr=Si double bond has no rotational barrier worth mentioning. This applies also, with some restrictions, to the discussed base adducts.

The ^{13}C-NMR spectra of 4–7, 9–11 show a close similarity to the spectral data of analogous carbene complexes. The shift differences between the metal carbonyls of the silylene complexes and the related carbon compounds are only small. These results underline the close analogy between the silicon compounds 4–7, 9–11 and Fischer carbene complexes. This view is also supported by the IR spectral data. On the basis of an analysis of the force constants of the v_{CO} stretching vibration,

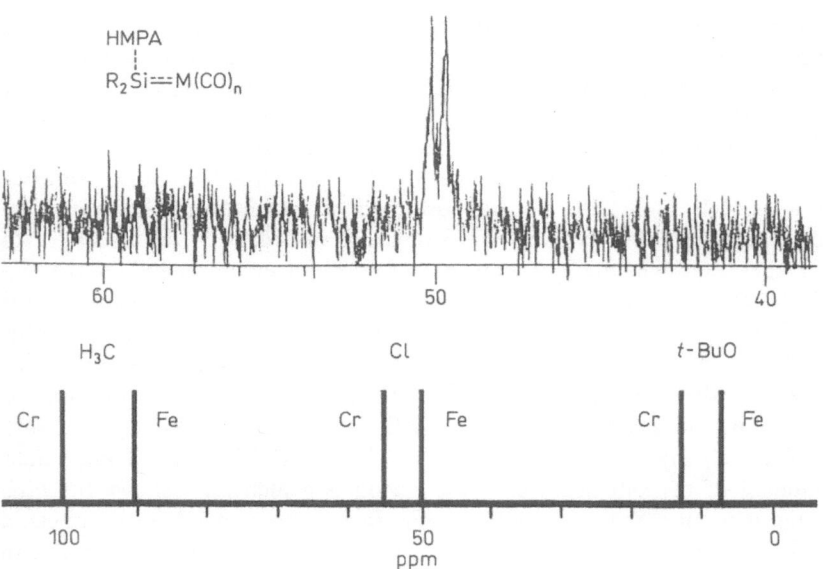

Fig. 4. ^{29}Si-NMR data of selected silylene complexes (ref. to table 1)

Table 2. IR data of silylene complexes [36]

Compound	$\nu \, CO/cm^{-1}$	k/Nm^{-1}
$(OC)_4FeSi(HMPA)(t\text{-}BuO)_2$	2005, 1920, 1883	1623, 1489, 1432
$(OC)_5CrSi(HMPA)(t\text{-}BuO)_2$	2015, 1991, 1930	1640, 1601, 1504

a higher donor effect for the $Cr(CO)_5$ fragment compared to the $Fe(CO)_4$ fragment can be derived (Table 2). This interpretation is in accordance with a comparison of the MSi bond distances obtained by x-ray structure analysis.

2.2.2 Mössbauer Spectroscopy

Some surprising results were obtained from the Mössbauer spectra of iron silylene coordination compounds (Table 3). The observed quadrupole splitting of **4** is typical for a d^8 iron atom with trigonal bipyramidal coordination geometry. The observed isomer shifts, however, are unique in the literature, since the shift values are much more negative than in any other observed case (-0.477 **4**, -0.488 **22**) (Fig. 5). Presumably this effect is related to the axial coordination of the silicon and the strong acceptor capabilities of the metal fragment at this coordination site [144]. However, in order to give a more precise explanation, further experiments are necessary.

From these parameters, a high charge density on the iron nucleus can be inferred. It is interesting to note that this situation is not reflected by the spectroscopic

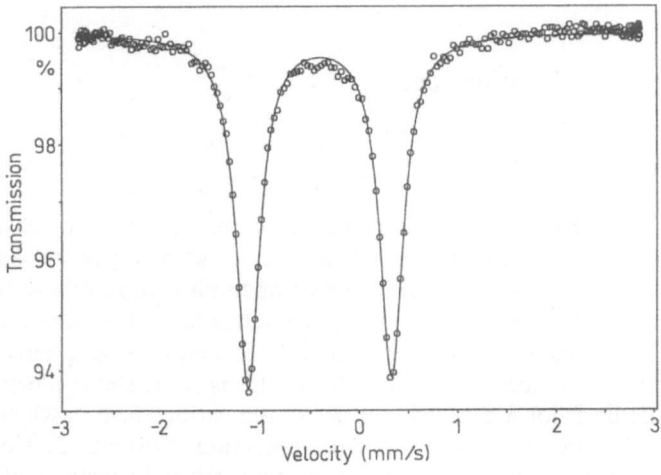

Fig. 5. Mössbauer spectrum of **4** at 4.2 K, source $^{57}Co:Rh$, Fe metal as standard

Christian Zybill

Table 3. Mössbauer data of iron silylene complexes (Fe as standard)

Compound	IS/mms^{-1}	QS/mms^{-1}
$(OC)_4FeSi(HMPA)_2Fe(CO)_4$	−0.488	1.62
$(OC)_4FeSi(HMPA)(t\text{-BuO})_2$	−0.477	1.46
$Na_2Fe(CO)_4$	−0.328	[a, b]
$Fe(CO)_5$	−0.178	[a, c]

[a] Goldanski VI, Herber RH (eds) (1968) Chemical applications of Mössbauer spectroscopy, Academic Press, New York
[b] T = 298 K, [c] T = 78 K

data of the substituents at the iron. A correlation between isomer shift and oxidation state of the metal need not necessarily exist; but a high charge density on the metal can be explained by a significant contribution of the formula **B**. This interpretation is also consistent with the data obtained by x-ray structure analysis.

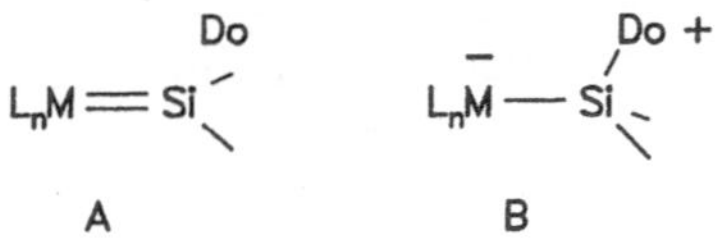

$$L_nM = Si \qquad L_nM - Si \qquad (15a)$$

A   B

The assignment of oxidation states has more a formal character in the sense of electron counting rules [145]. In this context it should, however, be justified to use at least the term low valent silicon.

2.3 X-Ray Structure Determinations

2.3.1 Structures of Silylene Complexes

Since 1987, a sufficient set of structural data of base-stabilized silylene complexes has been compiled to allow the deduction of general rules with respect to the bonding parameters at silicon. All the investigated compounds feature relatively short metal-silicon bonds together with a distorted tetrahedral coordination geometry. The significant deviations from the tetrahedral standard are a sensitive probe for the particular substituent effects at silicon, inclusive of steric effects. These data provide the basis for a detailed analysis of the various factors which have a predominant influence on the metal-silicon interaction. However, first of all some introductory remarks shall be made concerning the metal-silicon single bond.

20

Table 4. Selected structural data of silylene complexes, substituent effects and pyramidalization at Si [35–38]

Compound		M-Si/Å	Si–O/Å	M–CO/Å	C–O/Å	$\Sigma/°$
$(OC)_4FeSi(HMPA)(t\text{-}BuO)_2$	**4**	2.289(2)	1.730(3)	1.758(6)	1.156(6)	342.1
$(OC)_4FeSi(HMPA)(t\text{-}BuS)_2$	**5**	2.278(1)	1.734(2)	1.780(3)	1.151(3)	342.6
$(OC)_4FeSi(HMPA)Me_2$	**6**	2.280(1)	1.735(3)	1.792(6)	1.137(6)	339.0
		2.294(1)	1.731(3)	1.783(5)	1.147(5)	339.8
$(OC)_4FeSi(HMPA)Cl_2$	**7**	2.214(1)	1.683(3)	1.793(5)	1.143(5)	336.3
		2.221(1)	1.676(3)	1.796(5)	1.127(5)	337.7
$(OC)_5CrSi(HMPA)(t\text{-}BuO)_2$	**9**	2.431(1)	1.736(2)	1.855(3)	1.154(3)	342.2
$(OC)_5CrSi(HMPA)Me_2$	**10**	2.410(1)	1.743(2)	1.866(3)	1.148(4)	339.5
$(OC)_5CrSi(HMPA)Cl_2$	**11**	2.343(1)	1.690(3)	1.862(4)	1.156(5)	337.6

	M–Si/Å	Si–N/Å	$\Sigma/°$
$[Cp^*Ru(PMe_3)_2Si(CH_3CN)Ph_2]^+ \cdot$ $\cdot BPh_4^-$ **12**	2.328(2)	1.932(8)	351.6

Compared to the sum of covalent radii, metal-silicon single bonds are significantly shortened. This phenomenon is explained by a partial multiple bonding between the metal and silicon [62]. A comparison of several metal complexes throughout the periodic table shows that the largest effects occur with the heaviest metals. However, conclusions drawn concerning the thermodynamic stability of the respective M – Si bonds should be considered with some reservation [146], since in most cases the compared metals show neither the same coordination geometries nor the same oxidation states.

On the other side, only a relatively small shortening has to be expected for a metal-silicon double bond compared to a single bond. Accordingly, the observed MSi bond distances of the silylene complexes are 2.431(1) **9**, 2.410(1) **10**, 2.343(1) **11**, and 2.289(2) **4**, 2.278(1) **5**, 2.280(1)/2.294(1) **6** and 2.214(1)/2.221(1) Å **7** (Table 4). A simple addition of the covalent radii of Fe, Cr, and Si of the hypothetical compounds $(OC)_4Fe = SiMes_2$ gives FeSi 2.18 Å and for $(OC)_4Cr = SiMes_2$, CrSi 2.28 Å; the covalent radii for the single bonds are FeSi 2.6 Å, CrSi 2.7 Å. An ab-initio calculation for $(OC)_5Cr = Si(OH)H$ yields 2.4 Å for an uncomplexed Cr = Si double bond.

Both series of Fe and Cr complexes show a characteristic dependence on the respective donor and acceptor properties of the ligands at the silicon atom. These effects are more pronounced in the case of the chromium compounds and must be interpreted in terms of a higher degree of multiple bonding between Cr and Si. Generally, the bond distances can be correlated with the donor capabilities of the substituents at the silicon. A reduced donor capacity of the substituents leads to a more electrophilic silicon and thus to a shortening of the metal silicon bond, together with a stronger coordination of the HMPA donor. The stronger

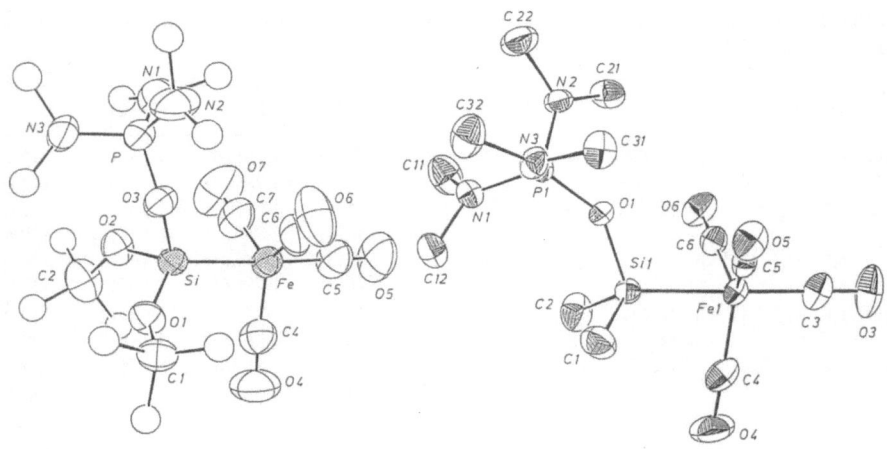

Fig. 6. Coordination geometry at silicon: structures of the complexes **4** (*left*) and **5** (*right*)

coordination of the HMPA is accompanied by an increased pyramidalization at the silicon atom (Figs. 6, 7; Table 4).

All characteristic distortions from the ideal tetrahedral geometry at silicon (large angles $M-Si-R$, small angles $R-Si-R$) are comparable to the ones observed in donor adducts of silaethenes and silaimines. A general interdependancy of the substituent effects between the various substituents at the silicon is obvious. In principle, the structures **4 − 7, 9 − 11** can be considered as models for the stepwise addition of a donor molecule to a trigonal planar silicon atom. The resulting pictures are like snapshots taken along a presumed reaction coordinate.

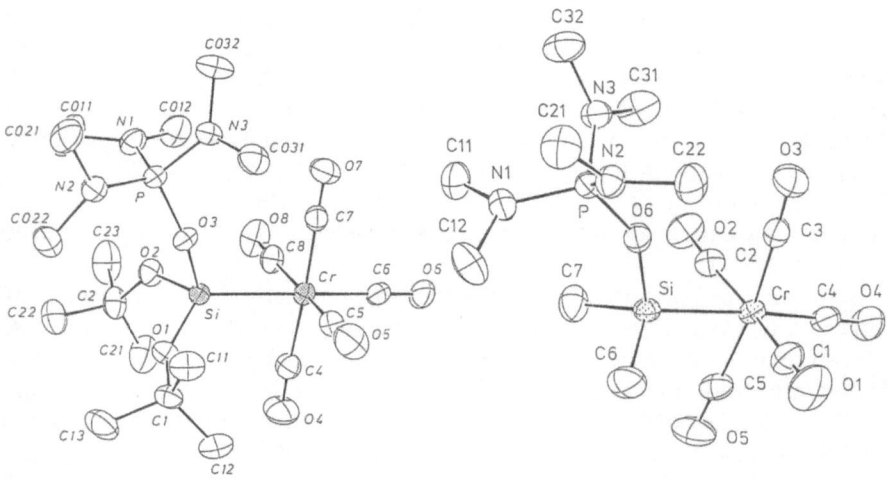

Fig. 7. Coordination geometry at silicon: structures of the compounds **9** (*left*) and **10** (*right*)

Finally, an example of an x-ray structure of a cationic complex shall be mentioned. From the data for **12**, a surprisingly weak coordination (Si—N 1.932(8) Å [146, 147]) of the acetonitrile donor to the silicon is inferred. The deviation from a pure tetrahedral geometry at the silicon is the largest yet observed (Table 4).

2.4 Theoretical Description of Base-Stabilized Silylene Complexes

For a theoretical consideration of the metal-silicon interaction in silylene complexes, the fragment orbital description proves to be very useful [148]. This approach has been extensively used in the organometallic chemistry of carbon and allows a basic understanding of the interrelations also by means of a qualitative description.

The required fragment orbitals for C_{3v} Fe(CO)$_4$ and C_{4v} Cr(CO)$_5$ are taken from the literature, also the orbitals of the respective silylenes (Fig. 8).

A Cr=Si double bond is essentially composed of two components; an electron donating σ-interaction between the a_1 (3s) orbital at the Si and the empty a_1 (d_z^2) orbital at Cr and a π-back bond between the e_s (d_{yz}, d_{xz}) orbitals of Cr and the LUMO b_1 (p_y) at Si (Fig. 9).

An ab-initio calculation for (OC)$_5$Cr=Si(OH)H shows parallels to the parent carbon compound. However, the electron deficiency at the silicon atom is significantly higher compared to the carbene complex. The LUMO of the Cr=Si

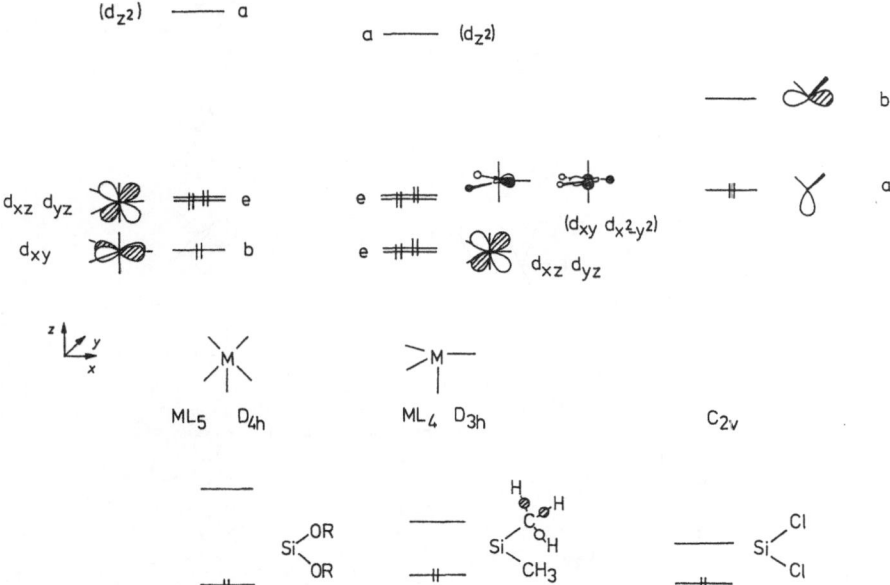

Fig. 8. Energy scheme of the frontier orbitals of the Cr(CO)$_5$- and Fe(CO)$_4$-fragments

double bond is π-antibonding and has a maximum coefficient of 0.85 at silicon with an orbital energy of 2.12 eV, whereas the coefficient for $(OC)_5Cr=C(OH)H$ is only 0.66 with an orbital energy of 3.86 eV. Therefore, the silylene complex should be more reactive towards nucleophiles than the carbene complex.

Since the LUMO is π-antibonding, the kinetically controlled interaction of a donor molecule (HMPA) with the silicon leads to a decrease of multiple bonding between Cr and Si concomitant with a pyramidalization at the silicon atom. The resulting MSi bond distance and pyramidalization effect are strongly influenced by the respective substituents.

An analysis of the bonding situation in a trigonal bipyramidal d^8 system shows that two different coordination sites (equatorial, apical) with different donor-acceptor capabilities are available (Fig. 8) [149, 150].

Ligands with a relatively high σ-donor and weak π-acceptor capacity prefer the apical coordination site, whereas the equatorial coordination site is occupied by strong π-acceptors. All considered iron complexes feature the silylene ligands at the axial position, which underlines the strong σ-donor and weak π-acceptor behavior of the base-stabilized silylenes. In the case of the chromium compounds, in particular the MSi bond distances indicate a much stronger metal-silicon interaction. The reason is a better $d_\pi p_\pi$ MSi back bond and the consequent higher degree of multiple bonding at chromium compared to the axial coordination site at the iron complex. This interpretation also seems plausible in the light of data

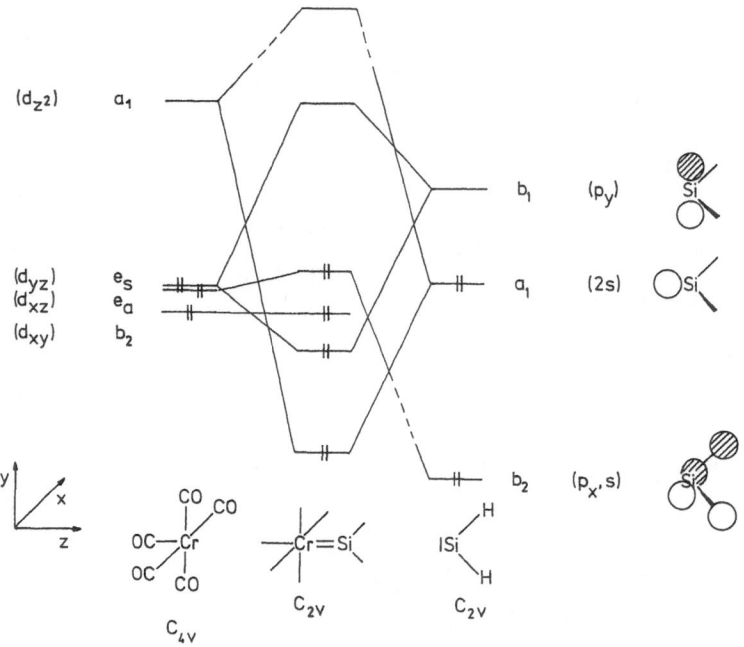

Fig. 9. Schematic representation of the Cr=Si bond in $(OC)_5Cr=Si(OH)H$

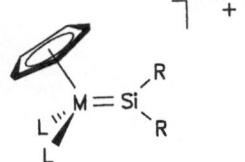

Fig. 10. HOMO of a d^6 CpML$_2$ fragment, which leads to the depicted preferred conformation of a silylene complex

from photoelectron spectroscopy. According to the first ionization energies of the metal carbonyls Cr(CO)$_6$ (8.41 eV for the ionization from the t_{2g}-HOMO) and Fe(CO)$_5$ (8.60, 9.90 eV for the ionization from the two degenerate sets of e orbitals), a higher donor capacity of the Cr(CO)$_5$ fragment can be derived [151–154].

A theoretical basis for the description of the cationic complex [Cp*Ru(PR$_3$)$_2$= =SiR$_2$]$^+$ can also be given. For a d^6 CpML$_2^+$ system, a complete splitting of the three t_{2g} orbitals (octahedral symmetry) is to be expected. Consequently, a coordinated silylene ligand (without any base) should prefer the indicated (Fig. 10) conformation.

2.5 Reactions of Silylene Complexes

2.5.1 The Reactivity of Stable Donor Adducts

Investigations of the reactivity of stable silylene complexes are still at an early stage of development. The reactions known so far, however, are of interest, since most of them are model cases which have important mechanistic implications.

A general reaction is the photochemical or base-induced cleavage of silylenes from the metal. The cophotolysis of 4 with a phosphine base first yields the *trans*-phosphine silylene complex 43. In a second step, the silylene is cleaved off the metal and the *trans*-bisphosphine complex 44 is formed. Presumably the initial step of this reaction is the liberation of the *trans*-CO with formation of an intermediate 16e species. An indication for this mechanism is the fact that the reaction can be suppressed by an excess of CO. The intermediate 16e complex is stabilized by the coordination of a phosphine. At this stage, the photochemical cleavage of the silylene occurs. Again, a reactive 16e species is formed which takes up a further phosphine ligand. If no trapping reagents are present, the silylenes form chain polymers of low molecular weight, which can be identified by NMR and mass spectroscopy (eq. 16).

An oxygen transfer from HMPA-O to silicon, which is known from the reduction of phosphine oxides with chlorosilanes, is not observed. On the contrary, the liberated HMPA can be recovered quantitatively (eq. 17, 18).

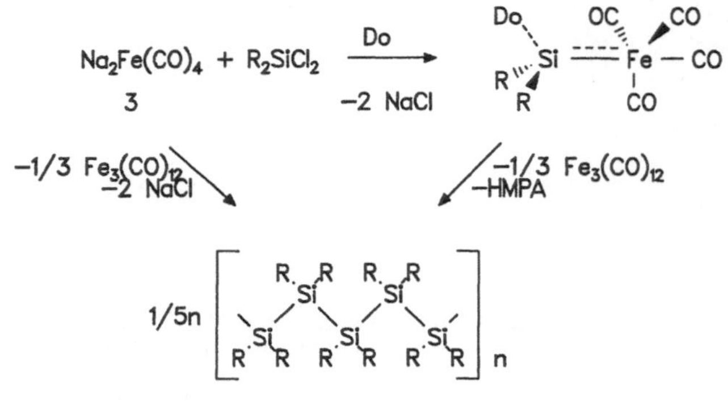

$$Na_2Fe(CO)_4 + R_2SiCl_2 \xrightarrow[-2\ NaCl]{Do}$$

(with products shown) ... (16)

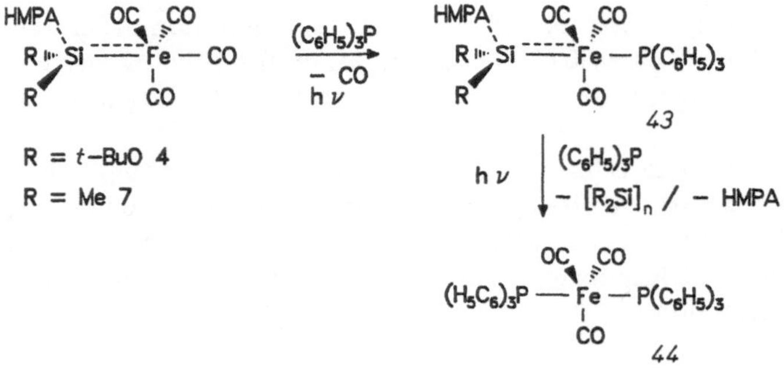

$$-1/3\ Fe_3(CO)_{12}\ -2\ NaCl$$

$$1/5n \left[\begin{array}{c} R\ R\ R\ R \\ Si\ Si \\ Si\ Si\ Si \\ R\ R\ R\ R\ R\ R \end{array} \right]_n$$

R = t–BuO 4
R = Me 7

$$\xrightarrow[h\nu]{(C_6H_5)_3P,\ -CO}$$

43

$$h\nu \downarrow (C_6H_5)_3P,\ -[R_2Si]_n\ /\ -HMPA$$ (17)

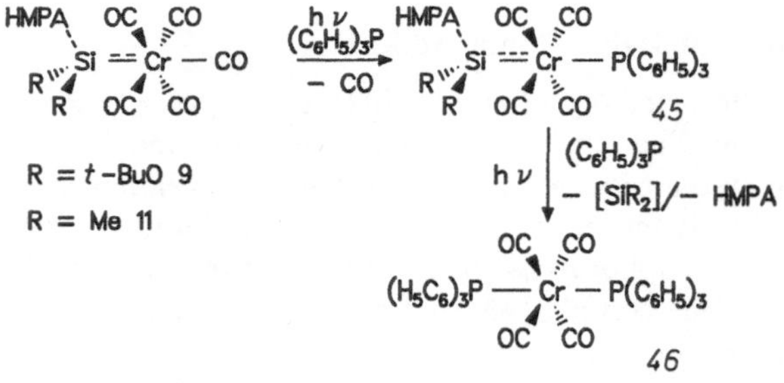

$$(H_5C_6)_3P - Fe - P(C_6H_5)_3$$

44

R = t–BuO 9
R = Me 11

$$\xrightarrow[-CO]{h\nu,\ (C_6H_5)_3P}$$

45

$$h\nu \downarrow (C_6H_5)_3P,\ -[SiR_2]/-HMPA$$ (18)

$$(H_5C_6)_3P - Cr - P(C_6H_5)_3$$

46

Recently, a variety of silylenes were generated and characterized by matrix isolation techniques. The observed loose donor adducts between silylenes and the matrix molecules (THF, CO) are only stable at very low temperatures. Melting of the matrix induces polymerization of the silylenes which proceeds through disilenes. However, O→Si transfer reactions do not occur; only in the case of 1-methyl-THF has an insertion of the silylene into the C−O bond been observed [155–158].

The photolysis of 4 in the presence of 2,3-dimethylbutadiene yields 3,4-dimethyl-1-silacyclobut(3)enes 47a,b as the reaction product of a formal [4 + 1] cycloaddition reaction. These results indicate a cleavage mechanism of the silylene from the metal concomitant with the liberation of HMPA and a subsequent [4 + 1] cycloaddition of the silylene to the butadiene. Prolonged photolysis in an excess of butadiene also yields the butadieneirontricarbonyl complex [159] (eq. 19).

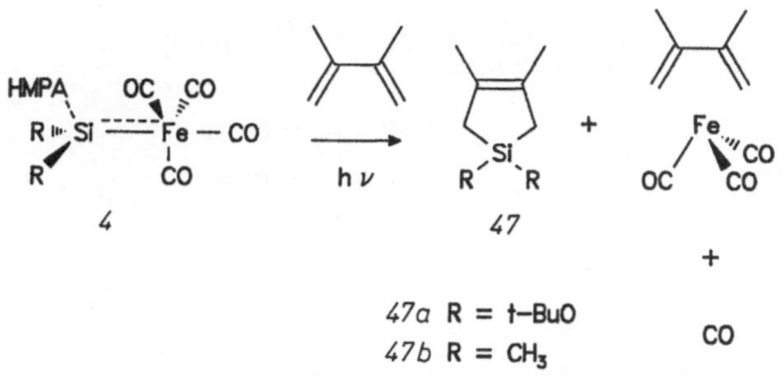

$$47a \ R = t\text{–}BuO$$
$$47b \ R = CH_3$$

(19)

At this stage of the discussion it is obvious that stable donor adducts of silylene complexes show a modified silylene reactivity and can thus be considered as model compounds for otherwise inaccessible reactive intermediates.

The activation of silylene complexes is induced both photochemically or by addition of a base, e.g. pyridine. A similar base-induced cleavage is known from the chemistry of carbene complexes; however, in this case the carbenes so formed dimerize to give alkenes. Finally, a silylene cleavage can also be achieved thermally. Melting of the compounds 4–7 in high vacuum yields the dimeric complexes 48–51 with loss of HMPA. The dimers, on the other hand, can be transformed into polysilanes and iron carbonyl clusters above 120 °C. In all cases, the resulting polymers have been identified by spectroscopic methods.

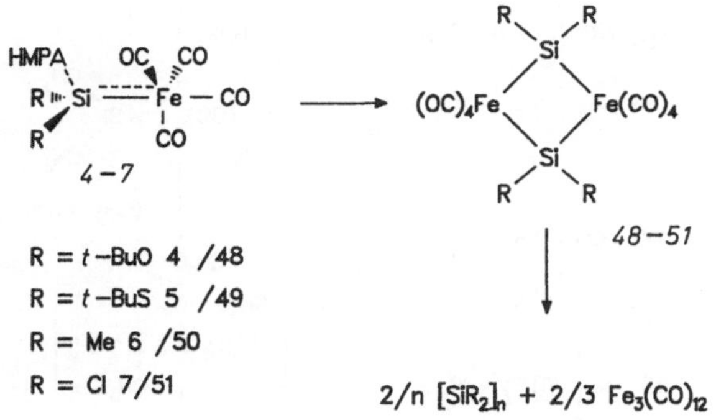

$$R = t\text{–}BuO \ 4 \ /48$$
$$R = t\text{–}BuS \ 5 \ /49$$
$$R = Me \ 6 \ /50$$
$$R = Cl \ 7/51$$

(20)

$$2/n \ [SiR_2]_n + 2/3 \ Fe_3(CO)_{12}$$

2.5.2 Reactions of Chloro-Substituted Silylene Complexes

Stable silylene complexes with halogen substituents are a useful starting material for further displacement reactions. Starting from 7 and 11, the Si complexes 22–24 are accessible in high yield. However, the intermediate dichlorosilylene complex

Christian Zybill

does not need to be isolated since the reaction can be performed in a one pot procedure simply by mixing the carbonyl metalate and $SiCl_4$ in a 2:1 ratio.

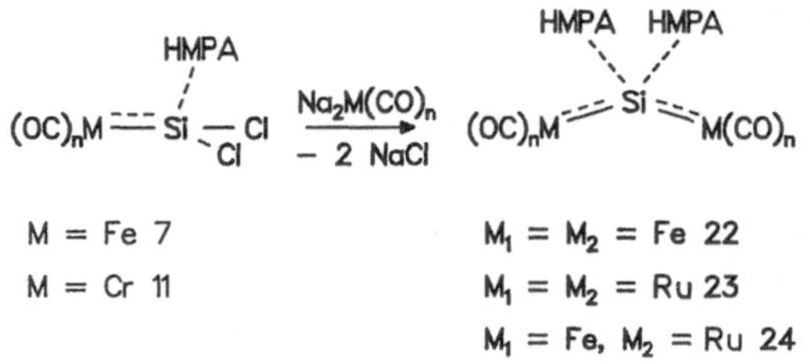

$$
\underset{M = Fe\ 7}{\underset{M = Cr\ 11}{(OC)_nM \overset{\overset{\displaystyle HMPA}{\diagup}}{=\!=\!=} Si \overset{Cl}{\underset{Cl}{\diagdown}}}} \quad \xrightarrow[-\ 2\ NaCl]{Na_2M(CO)_n} \quad \underset{\substack{M_1 = M_2 = Fe\ 22 \\ M_1 = M_2 = Ru\ 23 \\ M_1 = Fe,\ M_2 = Ru\ 24}}{(OC)_nM \overset{\overset{\displaystyle HMPA\ \ HMPA}{\diagup\ \ \diagup}}{=\!=} Si \overset{}{=\!=} M(CO)_n}
$$

(21)

2.5.3 Silylene Complexes as Reactive Intermediates

"Base-free" uncomplexed silylene complexes are so far only known as reactive intermediates which are generated at low temperatures and trapped by suitable reagents. Several publications related to this subject are known, but most of the work is now summarized in review articles [95].

Also, reactive silylene complexes of iron and chromium can be generated at low temperatures and subsequently derivatized by trapping reagents. In THF as solvent, first labile THF adducts are formed, which are converted to the more stable HMPA adducts. The THF complexes dimerize above $-40\,°C$ with loss of THF. The silylene complexes can be utilized for reactions if they are generated in the presence of reagents like dimethylcarbonate. The resulting reaction products

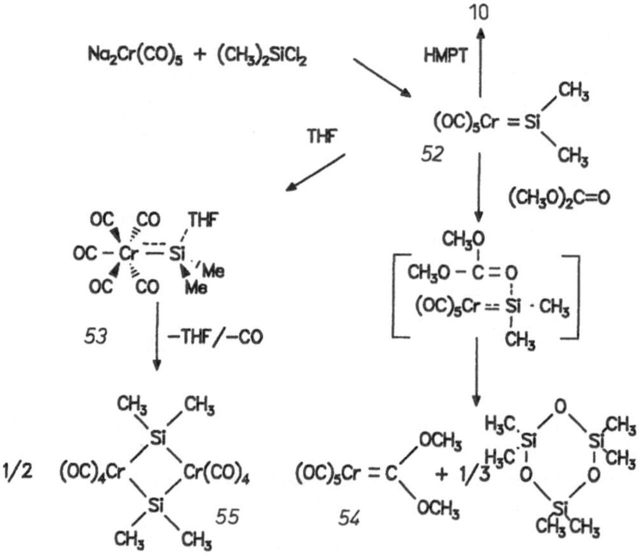

(22)

28

can be understood in terms of a formal [2 + 2] cycloaddition reaction, followed by a cycloreversion process. However, it is rather likely that the reaction mechanism is not a concerted one but proceeds through polar intermediates.

As already mentioned, the silylenes can be photolytically cleaved from the metal and trapped by 2,3-dimethylbutadiene [159]. Particularly interesting are trapping experiments of silylenes which have been generated directly in the coordination sphere of the metal. The photolysis of **36** in the presence of triethylsilane yields the trapping product **58**. Similar experiments can be performed with PhSi(OMe)$_3$, PhCH$_2$OH, dienes, and ketones. On the basis of these experiments, a ferra-substituted silylene was unambiguously identified as reactive intermediate [157]. It is interesting to mention in this context related experiments involving the photochemical generation of coordinated phosphinidenes, for which similar precursors were used [160].

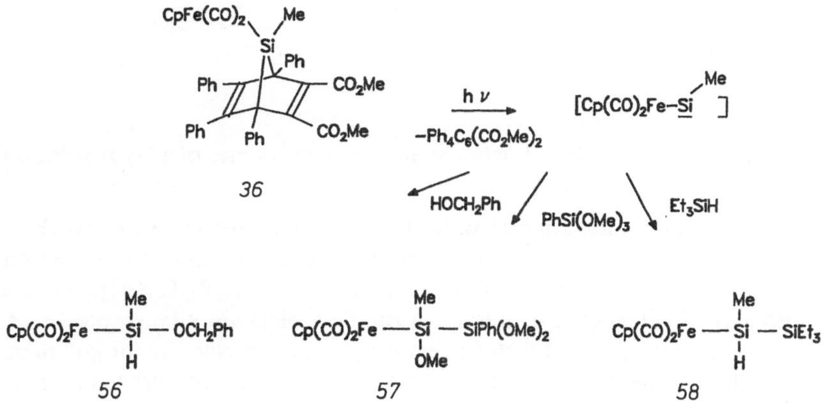

$$(23)$$

Cyclopentadienyldicarbonyldisilyliron complexes show very selective chemical reactivities. For instance, irradiation of **37** leads to a degradation of the silyl chain. Such deoligomerization reactions are known from the chemistry of polysilanes and proceed through silylenes as intermediates. In the above case, the reaction occurs in the coordination sphere of the metal. A prerequisite for the proposed mechanism is the assumption of a reversible CO exchange reaction. The resulting 16e species is stabilized by the migration of a silyl group to the metal. As by-products, polysilanes of unspecified chain length are formed by the poly-merization reaction of the liberated silylenes (Pannell rearrangement) [138, 139].

Similar results were obtained by Ogino and co-workers. After the initial cleavage of the CO from the metal, a silyl group migration occurs. It was shown by cross experiments that in this case the silyl substituent migrates and not the silylene. In a subsequent step, the silylene is then displaced by CO. Unfortunately, no trapping experiments of the silylenes in solution have been reported.

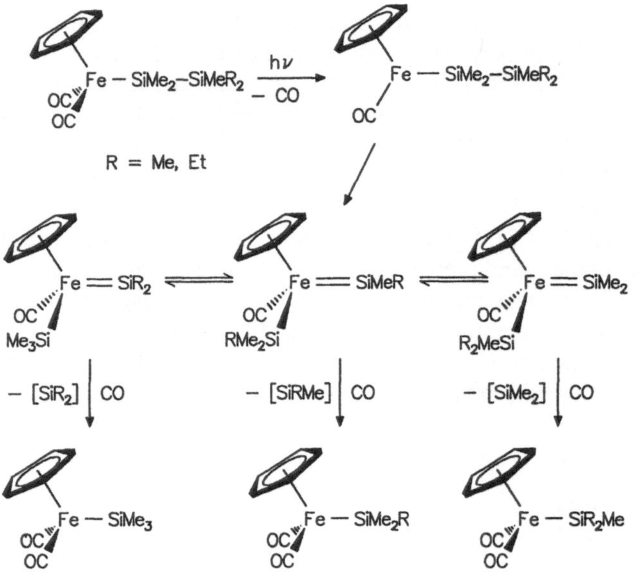

Christian Zybill

(24)

R = Me, Et

2.5.4 The Formation of SiSi Bonds from Silanes in the Presence of a Hydrosilation Catalyst

In 1971, a short communication was published [54] by Kumada and co-workers reporting the formation of di- and polysilanes from dihydrosilanes by the action of a platinum complex. Also the Wilkinson catalyst $(Ph_3P)_3RhCl$ promotes hydrosilation. If no alkenes are present, formation of chain silanes occurs. A thorough analysis of the product distribution shows a high preference for polymers (without a catalyst, disproportionation reactions of the silanes prevail). Cross experiments indicate the formation of a silylene complex as intermediate; and in solution, free silylenes could also be trapped by Et_3SiH [55, 56].

$$PhSiH_3 \xrightarrow[Et_3SiH]{(Ph_3P)_3RhCl} Ph_2SiH\text{-}SiEt_3 + Ph_2SiH\text{-}SiHPh_2 + Ph_3SiH \quad (25)$$
$$\qquad\qquad\qquad\qquad 59 \qquad\qquad\qquad 60$$

Recently, this work has been extended and further developed by Brown-Wensley into a preparative method for the synthesis of disilanes. The results of competitive reactions with several silanes allow insight into the reaction kinetics, in particular the relative rates of disilane formation versus hydrosilation (Table 5a, b) [61].

$$PhMeSiH_2 \xrightarrow{(Ph_3P)_2Pt(H_2C=CH_2)} PhMeSiH\text{-}SiHPhMe + Ph_2MeSiH + H_2$$
$$\qquad\qquad\qquad\qquad 61 \quad 27\% \qquad\qquad 52\% \qquad\qquad (26)$$
$$\qquad\qquad + \text{ others (including Pt compounds)}$$

30

Table 5a. Reaction of R_3SiH in the presence of $L_2Pt(C_2H_4)$[a]

Silane	$R_3Si-SiR_3$, %	Others, %
$PhMeSiH_2$	27	Ph_2MeSiH 52
$PhMe_2SiH$	0	Ph_2Me_2Si 4
Et_3SiH	0	—
Et_2SiH_2	95	—
$(EtO)_3SiHI$	0	$(EtO)_4Si$ 61
Et_2SiH_2 +	$Et_4Si_2H_2$ 58	
Et_3SiH	Et_5Si_2H 8	

[a] 0.03 mmol of catalyst, 9 mmol of silane

Table 5b. Relative rates of disilane formation and hydrosilation process in the presence of various catalysts

Catalyst	Si—Si bond formation[a]	Hydrosilation
$(Ph_3P)_2Pt(C_2H_4)$	1.0	0.8
H_2PtCl_6	0.1	6
$(Ph_3)_2PtCl_2$	0.1	0.1
$Pt(COD)Cl_2$	0.7	70
$(Ph_3P)_3RhCl$	31	400
$[Rh(CO)_2Cl]_2$	5	60
$(\eta^5-C_5H_5)Rh(C_2H_4)$	0.2	0.7
$RhCl_3$	0.3	3
$[Rh(COD(Cl]_2$	1	25
$[Ir(COD)Cl]_2$	0.2	3
$[Pd(allyl)Cl]_2$	12	100

[a] 0.01 mmol of catalyst and 1.0 mmol of Et_2SiH_2, the rate for $(Ph_2P)_2Pt(C_2H_4)$ was arbitrarily assigned as 1, for details see Ref. [61]

The mechanism of this reaction is still under discussion; however, it seems to be relatively clear that in this case silylene complexes are involved as intermediates.

The polymerization reaction of silanes with Cp_2ZrMe_2 as catalyst has also been investigated by several research groups. Some evidence for a reaction mechanism proceeding through silylene complexes as intermediates has been given

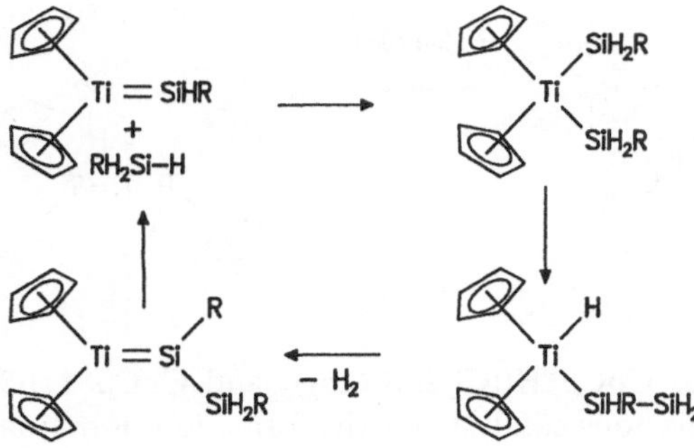

(27)

in the work of Corey [161, 162 a], but refer also to [162 b], for a MO calculation to [162 c].

In the case of the polymerization of *n*-butylsilane with Cp_2ZrMe_2, also a large percentage of cyclic Si_5 and Si_7 oligomers has been identified among the reaction products [163].

The most effective catalysts for the dehydrogenative coupling of silanes are the coordination compounds of the group IV metals such as Cp_2TiMe_2, Cp_2ZrMe_2, etc. In the case of the titanium catalysts, the resulting polysilanes were isolated and the molecular weight distribution ($M_w = 1500$) was measured by GP chromatography. From these reaction mixtures, two dimeric silylene complexes (**62** and **63**) could be isolated and characterized by x-ray structure analysis. On the basis of these experimental data, a mechanism is proposed which involves the oxidative addition of phenylsilane to **63**, followed by a subsequent silylene cleavage. This cycle involves only unambiguously identified compounds.

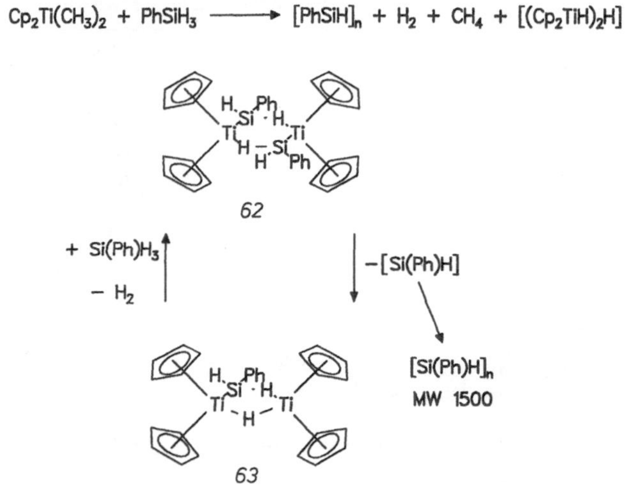

$$Cp_2Ti(CH_3)_2 + PhSiH_3 \longrightarrow [PhSiH]_n + H_2 + CH_4 + [(Cp_2TiH)_2H]$$

62

$+ Si(Ph)H_3$
$- H_2$

$-[Si(Ph)H]$

(28)

$[Si(Ph)H]_n$
MW 1500

63

Also the titanium hydride can react with $PhSiH_3$ to form **63**, followed by the extrusion of silylene again with formation of the hydride.

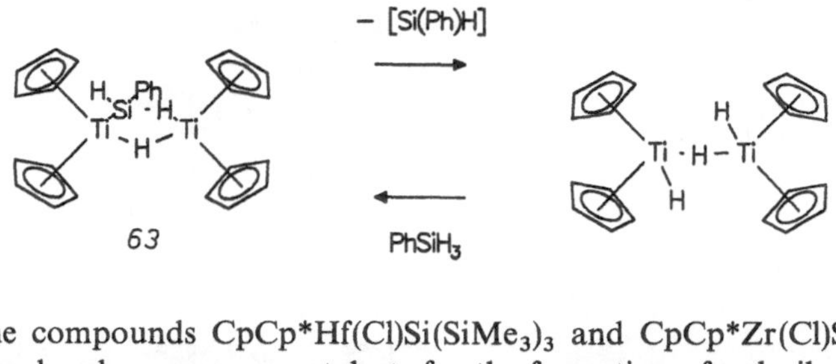

$- [Si(Ph)H]$

63          $PhSiH_3$

(29)

Recently, the compounds $CpCp^*Hf(Cl)Si(SiMe_3)_3$ and $CpCp^*Zr(Cl)Si(SiMe_3)_3$ have been used as homogenous catalysts for the formation of polysilanes.

A careful investigation of the reaction kinetics and detailed trapping experiments allow the conclusion that in this case a σ-bond metathesis reaction mechanism applies. The polymerization reaction of $PhSiH_3$ by $CpCp^*Hf(SiH_2Ph)Cl$ has been monitored by ^{1}H-NMR spectroscopy. The data $k(75\,°C) = 1.1(1) \times 10^{-4}\,M^{-1}\,s^{-1}$, $\Delta H = 19.5(2)$ kcal mol^{-1}, $\Delta S = -21(1)$ eu and $k_H/k_D = 2.9(2)$ (75 °C) are in good agreement with the proposed mechanism with a metallacycle as transition state [164].

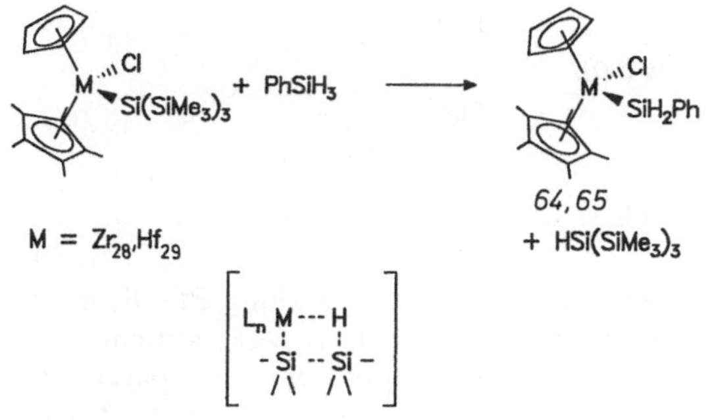

$$\tag{30}$$

$$M = Zr_{28}, Hf_{29}$$

64, 65
+ $HSi(SiMe_3)_3$

The discussion about the mechanism of the dehydrogenative polymerization reaction has not yet been completed. However, the reaction mechanism seems to be strongly influenced by the specific random conditions that apply for each particular system. Presumably with late transition metals a silylene mechanism is more appropriate. It may be a matter of the steric constraints of the system to shift the reaction towards σ-bond metathesis.

3 Dimeric Silylene Complexes

Bimetallic silylene-bridged complexes have been known for a long time and numerous articles related to this subject have appeared. Several compounds have been characterized, some of them also by x-ray structure analysis [165–171]. For instance, the complex $Mn_2(CO)_8(Si(C_6H_5)_2)_2$ shows a distorted $(MnSi)_2$ four-membered ring with a $Mn-Mn$ bond [169]. In the following section selected examples which have been described recently or are of particular interest in the present context will be discussed.

Monomeric base adducts of silylene complexes can be transformed into dimeric compounds at elevated temperatures with loss of the donor. This applies also to reactive donor-free compounds.

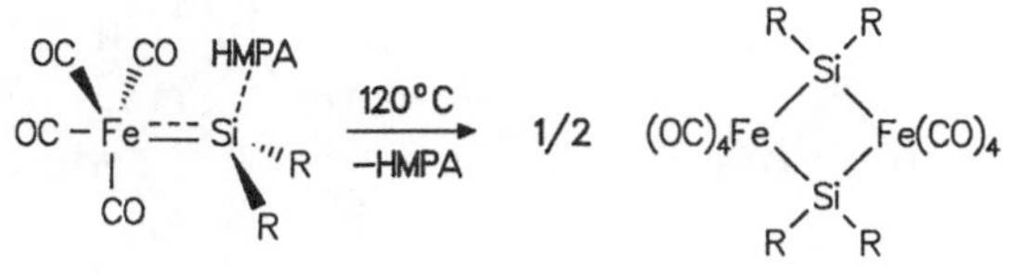

$$\tag{31}$$

It is interesting to note that no examples are known for a *retro*-reaction of this dimerization. Such a reaction has been observed, however, for germylene complexes; and for stannylene complexes, in some cases an equilibrium between uncomplexed and base-stabilized compounds has been found.

Of particular interest with respect to structure and bonding are the four-membered silaplatinum ring compounds **26abc**.

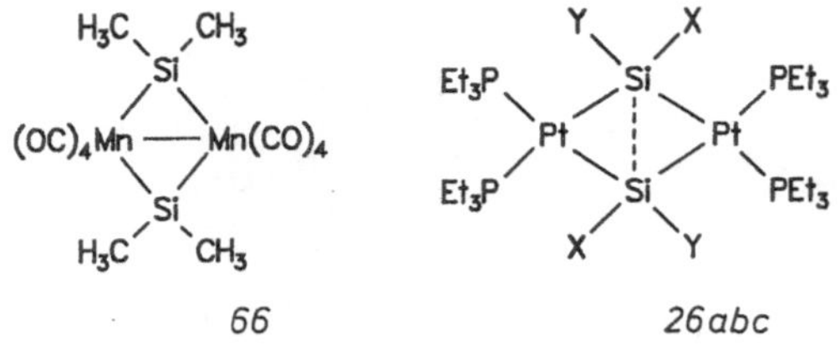

$$\text{(31a)}$$

In contrast to the above-mentioned Mn system, long Pt − Pt distances and very short cross ring SiSi distances were observed. An x-ray structure analysis of **1a–c** and **1b** shows a (PtSi)$_2$ four-membered ring with the parameters PtSi (1a–c 2.355(7)–2.383(8) Å), acute bond angles (Pt − Si − Pt 65.9(3)–66.6(1) °), and obtuse angles (Si − Pt − Si 113.6(1)–114.2(3) °). As a result of this diamond-shaped skeleton, the SiSi cross ring distance is 2.575(15)–2.602(4) Å, (van der Waals radius Si 1.17–1.18 Å, Pt 1.7–1.8 Å [172]) (The cross ring distance SiSi in **15** is 2.622(4) Å!). The authors definitely exclude repulsive van der Waals forces between the two Pt atoms (Pt − Pt distance 3.973(1)–3.997(2) Å). The crystallographic data are interpreted in the sense of a partial bonding interaction between the two silicon atoms. ASED-MO calculations support this interpretation [173], but refer also to the arguments of Schaefer and Grev concerning the "unsupported π-bond" in 1,2-disiladioxetanes [174].

Starting from (OC)$_5$MnSiR$_2$H (R = Me, Ph, Cl), the μ-silylene complex **70** is accessible via the oxidative addition of the Si − H bond to Pt(C$_2$H$_4$)(PPh$_3$)$_2$ and Pt(PPh$_3$)$_4$, respectively. Structure **70** can be functionalized by displacement of the phosphine ligands; alcoholysis and hydrolysis of the compound **70** leads to silicon-free complexes [175].

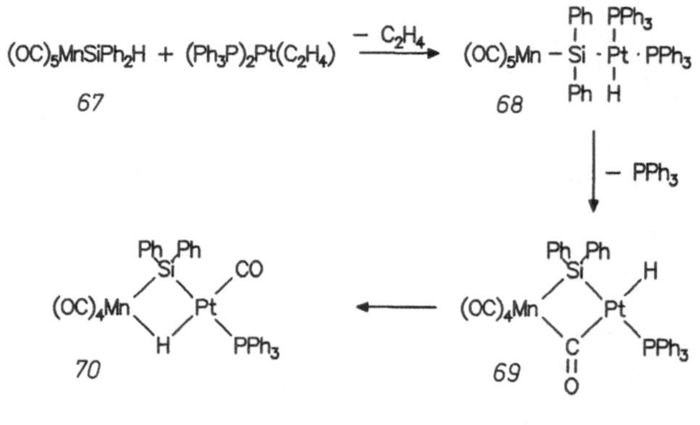

$$\text{(32)}$$

A photoinduced synthesis of the silylene-bridged dinuclear iron complex **71** has been accomplished from $CpFe(CO)_2SiMe_3$ and $RSiH_3$.

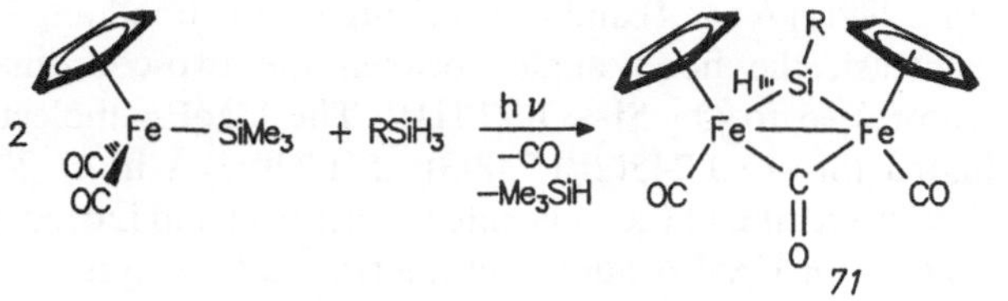

$$(33)$$

Pure *cis*-**71** could be isolated in 65% yield and was characterized by means of x-ray structure analysis (Fe−Si 2.270(1)/2.272(1) Å). This *cis*-conformer is exclusively formed. The formation of **71** requires a photoinduced silyl exchange followed by a decarbonylation reaction and further steps of an oxidative addition to the metal [176].

The chemistry of the μ-silylene bridge is still a relatively unexplored area, which is quite in contrast to the high degree of knowledge about its carbon congener, the μ-methylene bridge [117−183].

4 Coordination Compounds with Low Valent Silicon [184−189]

The chemistry of silicon in very low oxidation states is one of the most fascinating research areas, which can be located between molecular compounds of silicon and elemental (perhaps amorphous) silicon [190−194]. Most interesting results have recently been obtained by structural investigations of silicides in Zintl phases. However, compounds of silicon with negative oxidation states and very low coordination numbers of 1, 2, and 3 are so far only known in the composite of a crystal lattice.

Molecular compounds of silicon in a formally zerovalent oxidation state can be stabilized by appropriate transition-metal fragments. An entry to such polyme-tallated complexes of silicon is given by the chlorosilylene compounds **7**, **11** as a starting-point.

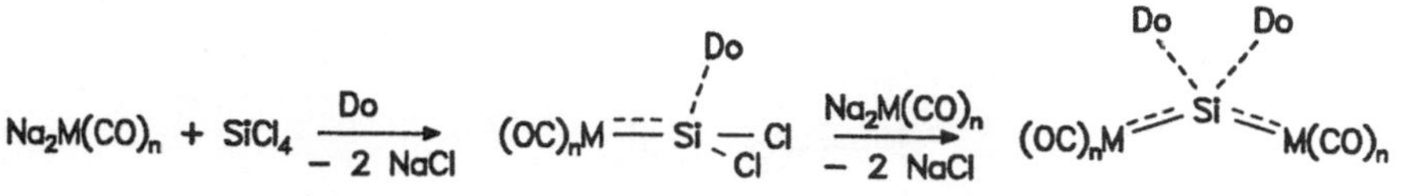

$$(34)$$

$M_1 = M_2 =$ Fe 22   **Do** = HMPA, pyr
$M_1 = M_2 =$ Ru 23
$M_1 =$ Fe, $M_2 =$ Ru 24

The combination of carbonylate dianions with silicon tetrachloride leads in high yields to the μ-Si compounds **22−24**. As already mentioned, the reaction can be performed either stepwise with isolation of the dichlorosilylene complex or in a one-pot procedure. The resulting products show a surprisingly high thermal

Christian Zybill

stability. Particularly interesting is the bimetallic compound **23**. All spectroscopic data of **22** are comparable with those of the base adducts of silylene complexes.

The x-ray structure of **22** features relatively long Fe—Si bonds (2.339(1), 2.341(1) Å, cf. on 2.289(2) Å for **4**) and a surprisingly wide bond angle Fe—Si—Fe 122.6(1)°. In contrast, the bond angle between the two-coordinated HMPA molecules is compressed to O—Si—O 92.1(1)°. The HMPA molecules are only loosely coordinated (Si—O 1.745(2)/1.748(3), cf. 1.730(3) Å in **4**). These features indicate a possible stretching of the molecule to linearity; and indeed, the structure can be interpreted as an HMPA adduct of a ferrasilaallene system (Fig. 11) [41].

A qualitative description of the bonding in the hypothetical complex $(OC)_4Fe = = Si = Fe(CO)_4$ can be derived by analogy to the situation of coordinated silylenes. With the assumption of two occupied sp-hybrid orbitals and two empty p-orbitals at the Si (0), two σ-donor and $d_\pi p_\pi$-backbonding interactions can be constructed. In close analogy to the silylene complex, two further donor molecules are coordinated to the silicon. Thus, the HMPA adducts can be interpreted as adducts of organometallic analogues of silaallene. The shown isolobal analogy supports this argument:

$$H_2C = Si = CH_2 \leftrightarrow (OC)_4Fe = Si = Fe(CO)_4$$

Supplementary to this subject, a recently published ab-initio calculation of silicondicarbonyl $O = C = Si = C = O$ should be mentioned [195]. With an extended basis set (DZP functions) a bent C_{2v} structure with C—Si—C 80.0°, Si—C 1.871, C—O 1.126 Å is calculated; the bent conformation is energetically more stable by 76.7 kcal/mol than a linear $D_{\infty h}$ structure [196]. These results also allow a reinterpretation of the experimental data (IR spectroscopy) for silicondicarbonyl [197].

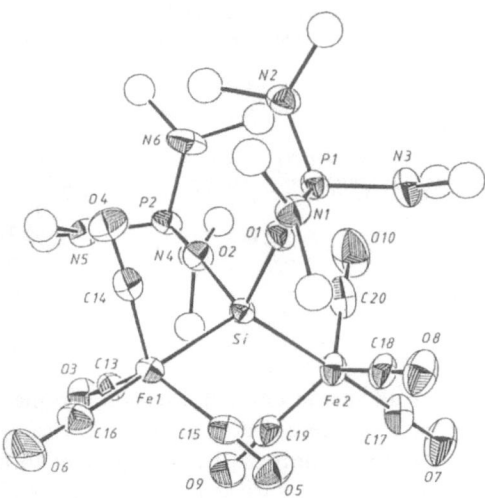

Fig. 11. Molecular structure of the complex bis-(tetracarbonyliron)silicon (0) × 2 HMPA **22**
$(OC)_4Fe = Si = Fe(CO)_4 \times 2$ HMPA

In the case of 2-silaallene, a bent structure also seems to be energetically favored.

Whether these results will also have an impact on the theory of metallaallenes is difficult to predict; at least for the compounds $Cp^*(CO)_2Mn = M = Mn(CO)_2Cp^*$, (M = Ge, Sn, Pb) a linear structure is established and also linear μ-carbido complexes are known [198]. Recently, a germanium compound has been synthesized which is directly comparable with 22. In this case, the starting material for the synthesis is not a monomeric base adduct, but a dimeric germylene complex which is cleaved by $Na_2Fe(CO)_4$ in pyridine to form 72 [199].

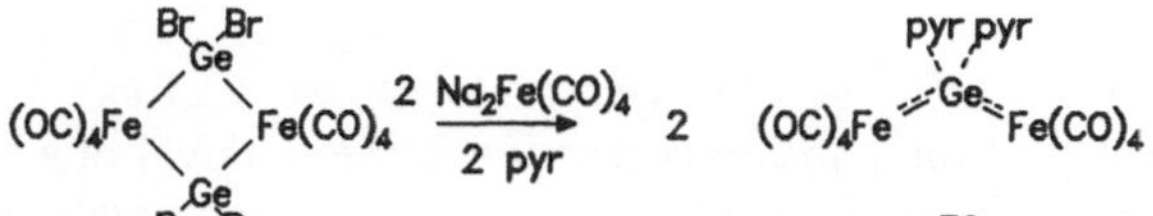

$$ (35) $$

The bond distances Fe – Ge 2.339(4), 2.351(3) Å are almost identical with the ones observed for 22 (2.339(1), 2.341(1) Å). The greater shortening of the Fe–Ge bond compared to silicon is due to the weaker donor capacity of the pyridine. Accordingly, the bond angle Fe – Si – Fe is widened to 131.0(1)°, whereas the two pyridine molecules still form almost a rectangle (89.4(6)°) to each other. Last but not least, an interesting reaction should be mentioned which gives an example for the functionalization of a germylene complex with Ge – Ge bond formation [199–201].

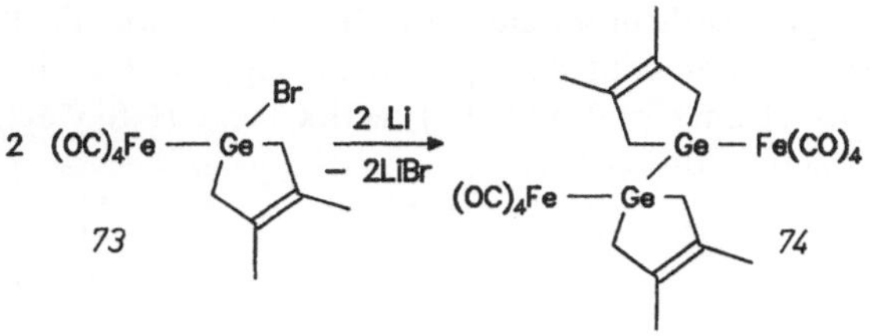

$$ (36) $$

The reaction of the perhalogenated compound yields only polymers.

5 η^2-Silene Complexes

During recent years, fascinating developments have occurred in the area of η^2-silene complexes, which opened up to totally new chemistry. Some of the highlights will be presented in this section. The first investigations of coordination compounds of silenes were carried out by means of matrix isolation techniques at very low temperatures. In particular, photochemical methods proved to be most effective for the generation of reactive η^2-silene complexes. Photolysis of the silylmethyl compound 75 provides the complex 76 almost quantitatively.

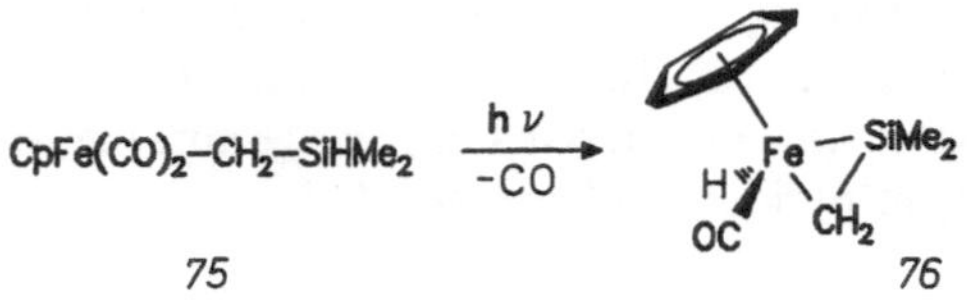

$$\text{CpFe(CO)}_2\text{-CH}_2\text{-SiHMe}_2 \xrightarrow[-\text{CO}]{h\nu}$$

(37)

75   76

The initiating step of the photolysis reaction is the removal of a CO ligand from the metal with generation of a reactive $16e$ species. The intermediate metal complex is stabilized by an intramolecular oxidative addition of the Si−H bond to the iron center.

This synthetic approach is known from the synthesis of $L_nM(\text{alkene})H$ compounds from $L_nM(\text{CO})$alkane precursors and can easily be applied to the analogous silyl complexes. The Si−H bond even shows an increased activity for oxidative addition reactions [42, 43].

The coordination compound 76 was stable enough for isolation and recording of its NMR spectra, from which a rigid silacyclopropane structure could be deduced. The mechanism of complex formation has also been investigated in detail by matrix techniques.

It appeared to be a logical consequence to transfer this synthetic principle to more suitable metals like ruthenium and introduce bulky, kinetically stabilizing ligands at the metal. An interesting example for this approach is the complex 78. The latter is synthesized from $\text{Cp*RuCl(PR}_3)_2$ with $\text{ClMgCH}_2\text{SiMe}_2\text{H}$ through 77 by a thermal Si−H activation reaction.

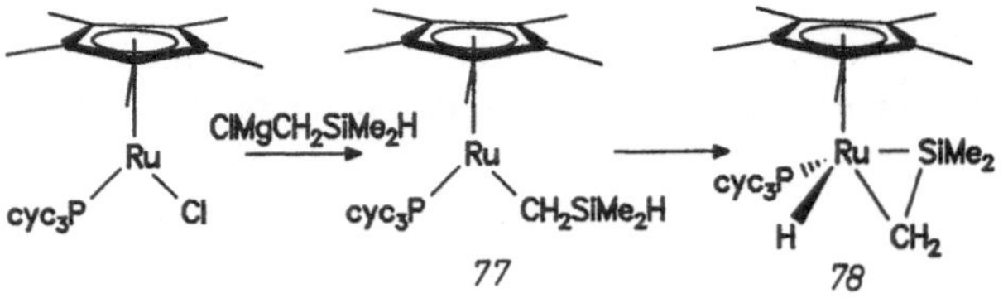

$$\xrightarrow{\text{ClMgCH}_2\text{SiMe}_2\text{H}} \longrightarrow$$

(38)

77   78

The x-ray structure of the stable compound 78 was determined. The bond distances are Ru−Si 2.382(4)/2.365(5) Å (2 enantiomers per asymmetric unit) and Si−C 1.78(2)/1.79(2) Å; which is consistent with the description of 78 as a ruthenasilacyclopropane (Fig. 12) [44].

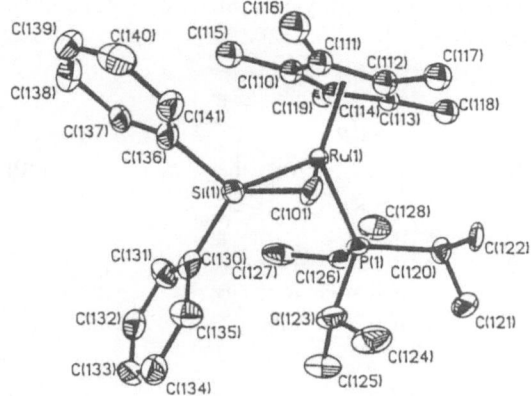

Fig. 12. ORTEP view of the silene complex **78** [44]

6 Coordination Compounds of Disilenes

The chemistry of the disilenes (disilaethenes) has developed very rapidly since the discovery of stable compounds. It was an obvious challenge to explore also the possibility of a π-coordination of disilenes to transition metals. According to the Dewar-Chatt-Duncason bonding model, a high stability for a disilene complex should result.

However, it was about 8 years after the first synthesis of tetramesityldisilene before stable coordination compounds became known. The main reason for this is the kinetic stabilization of the known disilenes by bulky substituents, which effectively prevents the coordination of the double bond to a metal fragment. Thus, a direct coordination of stable disilenes appeared to be reasonable only if metals with very low coordination numbers were used.

Evidence for a π-coordination was obtained through the reaction of various disilenes with $Hg(OCOCF_3)_2$, a reaction which leads regioselectively to bis(trifluoracetyl)disilanes. A disilene π-complex (**79**), which is stable up to $-50\,°C$, could be identified as an intermediate by spectroscopic methods.

$$RR'Si{=}SiR'R + Hg(OCOCF_3)_2 \longrightarrow \quad 79$$

$$\downarrow\ -Hg$$

$$RR'(CF_3COO)Si{-}Si(OCOCF_3)R'R$$

(39)

A particularly elegant pathway to stable coordination compounds of disilenes has been found with the reaction of the Pt-phosphine complex **80** with **81 ab**, which

Christian Zybill

gives the first stable disilene complexes **82ab** in high yields. Structures **82ab** have been characterized by NMR spectroscopy [47].

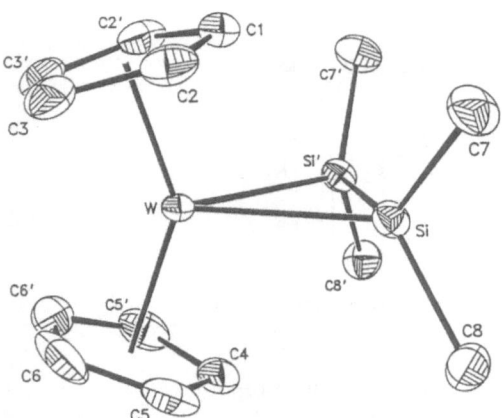

$$\begin{array}{ccc} \left[\begin{array}{c} R_2P \\ R_2P \end{array}\right. \!\!\!\! PtCl_2 + (R'_2SiH)_2 & \xrightarrow{Li} & \left[\begin{array}{c} R_2P \\ R_2P \end{array}\right. \!\!\!\! Pt \!\!\!\! \begin{array}{c} Si \\ Si \end{array} \end{array} \quad (40)$$

80   81a,b   82a,b

a R = Ph, R' = i–Pr
b R = R' = Ph

Very recently, synthesis and structure of molybdenum and tungsten complexes of the relatively unhindered disilene Si_2Me_4 were reported. The x-ray structure of **84** shows a metallacyclosilane structure with $W-Si = 2.606(2)$ Å and $Si-Si = 2.260(3)$ Å. The $W-Si$ bond length is within the range of various estimates of the Si and W covalent radii and the $Si-Si$ distance falls midway between the expected values for a single (2.35 Å) and a double bond (2.14 Å) (Fig. 13).

Structure **84** is the first example of a complex with small alkyl substituents at silicon and thus provides a valuable basis for the further development of the chemistry of disilaolefines [48].

$$\begin{array}{ccc} \text{83} & \xrightarrow{Mg} & \text{84} \end{array} \quad (41)$$

83   84

Fig. 13. ORTEP view of the disilene complex **84** [48]

7 Acknowledgement

It is a particular pleasure to thank the institutions which provided financial support to this work. These are the Deutsche Forschungsgemeinschaft, the Fonds der Chemischen Industrie, and the Bund der Freunde der Technischen Universität München. I am also indebted to Professor H. Schmidbaur, Garching, for his continued interest in our investigations, to Professor G. Müller, Konstanz, for the x-ray structure determinations and to Professor G. Bowmaker, who was on sabbatical leave from Auckland, New Zealand, for critical reading of the manuscript.

However, the work could not be done without the enthusiastic cooperation of all the people in our department, of which I should particularly mention Dipl. Chem. Chr. Leis and Dipl. Chem. H. Handwerker, who performed most of the experimental work in the laboratory.

Finally, I wish to thank my wife Helene and my family for their patience during the time this article was written.

8 References

1. West R (1987) Angew Chem 99: 1231; (1987) Angew Chem Int Ed Engl 26: 1201, further references therein
2. West R, Fink MJ, Michl J (1981) Science 214: 1343; (1984) Science 225: 1109 dd
3. Fink MJ, Michalczyk MJ, Haller KJ, West R, Michl J (1984) Organometallics 3: 793
4. West R (1984) Pure & Appl Chem 56: 163
5. Watanabe H, Okawa T, Kato M, Nagai Y (1983) J Chem Soc Chem Commun 781
6. Masamune S, Tobita H, Murakami S (1983) J Am Chem Soc 105: 6524
7. Schäfer A, Weidenbruch M, Pohl S (1985) J Organomet Chem 282: 305
8. Brook AG, Abdesaken F, Gutekunst B, Gutekunst G, Kallury RK (1981) J Chem Soc Chem Commun 191
9. Wiberg N, Wagner G, Reber G, Riede J, Müller G (1987) Organometallics 6: 35
10. Baines KM, Brook AG, Ford RR, Lickiss PD, Saxena AK, Chatterton WJ, Sawyer JF, Behnam BA (1989) Organometallics 8: 693
11. Müller G (1986) Nachr Chem Tech Lab 34: 778
12. Wiberg N, Schurz K, Fischer G (1985) Angew Chem 97: 1058; (1985) Angew Chem Int Ed Engl 24: 1053
13. Wiberg N, Schurz K, Reber G, Müller G (1986) J Chem Soc Chem Commun 591
14. Hesse M, Klingebiel U (1986) Angew Chem 98: 638; (1986) Angew Chem Int Ed Engl 25: 649
15. Smit CN, Lock FM, Bickelhaupt F (1984) Tetrahedron Lett 25: 3011
16. Collins S, Murakami S, Snow JT, Masamune S (1985) Tetrahedron Lett 26: 1281
17. Yoshifuji M, Shima I, Inamato N, Hirotsu K, Higuchi T (1981) J Am Chem Soc 103: 4587
18. Cowley AH, Lasch JG, Norman NC, Pakulski M (1983) J Am Chem Soc 105: 5506
19. Becker G (1976) Z Anorg Allg Chem 423: 242
20. Dimroth K, Hoffmann P (1964) Angew Chem 76: 433
21. Gier TE (1961) J Am Chem Soc 83: 1769
22. Märkl G, Sejpka H (1986) Tetrahedron Lett 27: 171
23. Yoshifuji M, Toyota K, Inamoto N (1984) J Chem Soc Chem Commun 689
24. Appel R, Fölling P, Josten B, Siray M, Winkhaus V, Knoch F (1984) Angew Chem 96: 620; (1984) Angew Chem Int Ed Engl 23: 619
25. Karsch HH, Köhler FH, Reisacher HU (1984) Tetrahedron Lett 25: 3687
26. Raabe G, Michl J (1985) Chem Rev 85: 419

27. Gaspar PP (1981) in: Jones M Jr, Moss RA (eds) Reactive Intermediates 2 Wiley, New York
28. Michalczyk MJ, Fink MJ, De Young DJ, Carlson CW, Welsh KM, West R, Michl J, Silicon J (1986) Germanium, Tin & Lead Compd 9: 1, 75
29. Raabe G, Vancik H, West R, Michl J (1986) J Am Chem Soc 108: 671
30. Lambert JB, McConnell JA, Schilf W, Schulz Jr, WJ(1988) J Chem Soc Commun 455
31. Maier G, Reisenauer HP, Schöttler K, Wessolek-Kraus U (1989) J Organomet Chem 366: 25
32. O'Neal HE, Ring MA, Richardson WH, Licciardi GF (1989) Organometallics 8: 1968
33. Walsh R(1989) Organometallics 8: 1973
34. Gordon MS, Boatz JA (1989) Organometallics 8: 1978
35. Zybill Chr, Müller G (1987) Angew Chem 99: 683; (1987) Angew Chem Int Ed Engl 26: 669
36. Zybill Chr, Müller G (1988) Organometallics 7: 1368
37. Straus DA, Tilley TD, Rheingold AL, Geib S (1987) J Am Chem Soc 109: 5872
38. Zybill Chr (1989) Nachr Chem Tech Lab 37: 248
39. Malisch W, Lorz P, Thum G (1988) Mainz, 13.–16.3, Abs. B 36
40. Ueno K, Tobita H, Shimoi M, Ogino H (1988) J Am Chem Soc 110: 4092
41. Zybill Chr, Wilkinson DL, Müller G (1988) Angew Chem 100: 574; (1988) Angew Chem Int Ed Engl 27: 583
42. Randolph CL, Wrighton MS (1987) Organometallics 6: 365
43. Lewis C, Wrighton MS (1983) J Am Chem Soc 105: 7768
44. Campion BK, Heyn RH, Tilley TD (1988) J Am Chem Soc 110: 7558
45. Zybill Chr, West R(1986) J Chem Soc Chem Commun 857
46. Pham EK, West R (1986) 20th Organosilicon Symposium Tarrytown, NY, April 18–19, P 2.3
47. Pham EK, West R (1989) J Am Chem Soc 111: 7667
48. Berry DH, Chey J, Zipin HS, Carroll PJ (1990) J Am Chem Soc 112: 452
49. Aylett BJ (1982) Adv Inorg Chem Radiochem 25: 1
50. Curtis MD, Epstein PS (1981) Adv Organomet Chem 19: 213
51. Aylett BJ (1980) J Organomet Chem Lib 9: 327
52. Cundy CS, Kingston BM, Lappert MF (1973) Adv Organomet Chem 11: 253
53. Büchner W (1980) J Organomet Chem Lib 9: 409
54. Yamamoto K, Okinoshima H, Kumada M (1971) J Organomet Chem 27: C 31 and references therein
55. Ojima I, Inaba SI, Kogure T, Nagai (1973) J Organomet Chem 55: C 7
56. Kumada M (1975) J Organomet Chem 100: 127
57. Seyferth D, Shannon ML, Vick SC, Lim TFO (1985) Organometallics 4: 57
58. Okinoshima H, Yamamoto K, Kumada M (1972) J Am Chem Soc 94: 9263
59. Sakurai H, Kamiyama Y, Nakadaira Y (1977) J Am Chem Soc 99: 3879
60. Pannell KH, Cervantes J, Hernandez C, Cassias J, Vincenti S (1986) Organometallics 5: 1056 and further references therein
61. Brown-Wensley KA (1987) Organometallics 6: 1590
62. Lappert MF, Maskell RK (1984) J Organomet Chem 264: 217
63. Speier JL (1977) Adv Organomet Chem 17: 407
64. Mackay KM, Nicholson BK in: Wilkinson G, Stone FGA, Abel EW (eds) Comprehensive Organometallic Chemistry, Pergamon Press 1982, and references therein
65. West R, David LD, Djurovich PI, Sinclair HYR (1983) Am Ceram Soc Bull 62: 899
66. Yajima S, Hayashi J Omori M, Okamura K (1976) Nature 261: 683
67. Yajima S, Omori M, Hayashi J, Okamura K, Matsuzawa T, Liaw C (1976) Chem Lett 551
68. West R, Wolff AR, Peterson DJ (1986) J Radiation Curing 13: 35
69. Miller RD, Hofer D, McKean DR, Willson CG, West R, Trefonas P (1984) in: Materials for Microlithography, Thomson L, Willson CG, Frechet JMJ (eds), ACS Symposium Series 266, American Chemical Society, Washington, DC

70. Zeigler JM, Harrah LA, Johnson AW (1985) SPIE Adv Resist Technol Proc II 539: 166
71. Hofer DC, Miller RD, Willson CG, Neureuther AR (1984) SPIE Adv Resist Technol Proc. 469: 16
72. Hench LL, Ulrich DR (eds) (1984) Ultrastructure processing of ceramics, glasses and composites, Wiley-Interscience, New York
73. Miller RD, Hofer D, Fickes GN, Willson CG, Marinero C, Trefonas III P, West R (1986) Polym Eng Sci 26: 1129
74. Trefonas III P, West R, Miller RD (1985) J Am Chem Soc 107: 2737
75. Harrah LA, Zeigler JM (1987) Macromolecules 20: 601
76. Kepler RG, Zeigler JM, Harrah LA, Kurtz SR (1983) Bull Am Phys Soc 28: 362
77. West R, David LD, Djurovich PI, Stearley KL, Srinivasan KSV, Yu H (1981) J Am Chem Soc 103: 7352
78. Yajima S (1983) Am Ceram Soc Bull 62: 899
79. West R, David LD, Djurovich PI, Yu H (1983) Am Ceram Soc Bull 62: 899
80. Schilling Jr CL, Wesson JP, Williams TC (1983) Am Ceram Soc Bull 63: 912
81. Kajzar F, Messier J, Rosilio C (1986) J Appl Phys 60: 3040
82. West R (1986) J Organomet Chem 300: 327
83. Miller RD (1989) Angew Chem Adv Mater 101: 1773
84. Nakatsuji H, Ushio J, Yonezawa T (1983) J Organomet Chem 258: C1
85. Nakatsuji H, Ushio J, Han S, Yonezawa T (1983) J Am Chem Soc 105: 426
86. Marks TJ, Seyam AM (1974) Inorg Chem 13: 1624
87. Schmid G, Balk HJ (1974) J Organomet Chem 80: 257
88. Thum G, Malisch W (1984) J Organomet Chem 264: C 5
89. Glockling F, Houston RE (1973) J Organomet Chem 50: C 31
90. Kang H, Jacobson DB, Shin SK, Beauchamp JL, Bowers MT (1986) J Am Chem Soc 108: 5668
91. Schmid G, Welz E (1977) Angew Chem 89: 823; (1977) Angew Chem Int Ed Engl 16: 785
92. The given value of the CN stretching vibration in the original paper [37] is incorrect
93. Hein F, Dobloth H (1941) Z Anorg Allg Chem 248: 84
94. Piper TS, Lemal D, Wilkinson G (1956) Naturwiss 43: 129
95. Tilley TD (1989) in: The chemistry of organic silicon compounds (Patai S, Rappoport Z, eds), Wiley
96. Collman JP, Finke RG, Cawse JN, Brauman JI (1977) J Am Chem Soc 99: 2515
97. Hieber W, Beck W, Braun G (1960) Angew Chem 72: 795
98. King RB (1964) Adv Organomet Chem 2: 157
99. Abel EW, Stone FGA (1969) Quart Rev 23: 325
100. Hieber W (1970) Adv Organomet Chem 8: 1
101. King RB (1970) Acc Chem Res 3: 417
102. Warnock GFP, Moodie LC, Ellis JE (1989) J Am Chem Soc 111: 2131
103. Zhen Y, Atwood JD (1989) J Am Chem Soc 111: 1506
104. Bockman TM, Kochi JK (1989) J Am Chem Soc 111: 4469
105. Griffith WP (1973) In: Comprehensive Inorganic Chemistry, Bailar JC, Emeleus HJ, Nyholm RS, Trotman DickensonAGF (eds), Pergamon Press, Oxford, vol. 4
106. Chi KM, Frerichs SR, Philson SB, Ellis JE (1987) Angew Chem 99: 1203; (1987) Angew Chem Int Ed Engl 26: 1190
107. Chi KM, Frerichs SR, Stein BK, Blackburn DW, Ellis JE (1988) J Am Chem Soc 110: 163
108. Schott G, Kibbel HU (1961) Z Anorg Allgem Chem 311: 53
109. Vancea L, Bennett MJ, Jones CE, Smith RA, Graham WAG (1977) Inorg Chem 16: 897; further references therein
110. Schmid G, Welz E (1979) Z Naturforsch 34b: 929
111. Jetz W, Graham WAG (1967) J Am Chem Soc 89: 2773
112. Leis Chr, Zybill Chr, Organometallics, submitted
113. Roddick DM, Heyn RH, Tilley TD (1989) Organometallics 8: 324; and references therein
114. Zarate EA, Tessier-Youngs C, Youngs WJ (1988) J Am Chem Soc 110: 4068
115. Chang LS, Johnson MP, Fink MJ (1989) Organometallics 8: 1369
116. Woo HG, Tilley TD (1989) J Am Chem Soc 111: 3757

117. Aitken CT, Harrod JF, Samuel E (1986) J Am Chem Soc 108: 4059
118. Aitken C, Harrod JF, Gill US (1987) Can J Chem 65: 1804
119. Aitken C, Harrod JF, Samuel E (1986) Can J Chem 64: 1677
120. Harrod JF (1988) In: Inorganic and organometallic polymers, Zeldin M, Wynn KJ, Alcock HR (eds) ACS Symposium Series 360, Washington DC
121. Speier JL (1979) Adv Organomet Chem 17: 407
122. Brunner H (1983) Angew Chem 95: 921; (1983) Angew Chem Int Ed Engl 22: 897
123. Cotton FA, Wilkinson G (1988) Advanced Inorganic Chemistry, Wiley, New York
124. Hart-Davis AJ, Graham WAG(1971) J Am Chem Soc 93: 4388
125. Dong DF, Hoyano JK, Graham WAG (1981) Can J Chem 59: 1455
126. Smith RA, Bennett MJ (1977) Acta Chryst B 33: 1113
127. Schubert U, Ackermann K, Wörle B (1982) J Am Chem Soc 104: 7378
128. Kraft G, Kalbas C, Schubert U (1985) J Organomet Chem 289: 247
129. Luo XL, Crabtree RH (1989) J Am Chem Soc 111: 2527
130. Schubert U, Müller J, Alt HG (1987) Organometallics 6: 469
131. Schubert U, Scholz G, Müller J, Ackermann YK, Wörle B, Stansfield RFD (1986) J Organomet Chem 306: 303
132. Lichtenberger DL, Chaudhuri AR (1989) J Am Chem Soc 111: 3583
133. see also: Rabaa H, Saillard JY, Schubert U (1987) J Organomet Chem 330: 397
134. Berry DH, Mitstifer JH (1987) J Am Chem Soc 109: 3777
135. Berry DH, Jiang Q (1987) J Am Chem Soc 109: 6210
136. Marinetti-Mignani A, West R (1987) Organometallics 6: 141
137. Couldwell MC, Simpson J, Robinson WT (1976) J Organomet Chem 107: 323
138. Tobita H, Ueno K, Ogino H (1986) Chem Lett 1777
139. Pannell KH, Rozell JM, Hernandez C (1989) J Am Chem Soc 111: 4482
140. Blinka TA, Helmer BJ, West R (1984) Adv Organomet Chem 23: 193
141. Kuroda M, Kabe Y, Hashimoto M, Masamune S (1988) Angew Chem 100: 1795; (1988) Angew Chem Int Ed Engl 27: 1795
142. Ernst CR, Spialter L, Buell GR, Wilhite DL (1974) J Am Chem Soc 96: 5375
143. Leis Chr (1988) Diplomarbeit, Technische Universität München
144. a) Pannell KH, Wu CC, Long GJ (1980) J Organomet Chem 186: 85
    b) Pebler J, Petz W (1977) Z. Naturforsch. 32 b: 1431
145. Green MLH, Green JC (1987) In: Seddon EA, Seddon KR (eds) The chemistry of ruthenium, Elsevier, Amsterdam
146. Tilley TD, Arnold J, Campion BK, Woo HG, Elsner F, Heyn RH, Rheingold Al, Geib SJ (1988) Third Chemical Congress of North America, 5–10 June, Toronto, Abstr. INOR 62
147. Zigler SS, Haller KJ, West R (1989) Organometallics 8: 1656
148. Albright TA (1982) Tetrahedron 38: 1339
149. Rossi AR, Hoffmann R (1975) Inorganic Chemistry 14: 365
150. Elian M, Hoffmann R (1975) Inorganic Chemistry 14: 1058
151. Böhm MC, Daub J, Gleiter R, Hofmann P, Lappert MF, Öfele K (1980) Chem Ber 113: 3629
152. Block TF, Fenske RF (1977) J Am Chem Soc 99: 4321
153. Johnson JB, Klemperer WG (1977) J Am Chem Soc 99: 7132
154. Hay PJ (1983) J Am Chem Soc 100: 2411
155. Gillette GR, Noren GH, West R (1989) Organometallics 8: 487
156. Akasaka T, Nagase S, Yabe A, Ando W (1988) J Am Chem Soc 110: 6270
157. Pearsall MA, West R (1988) J Am Chem Soc 110: 7228
158. Arrington CA, Petty JT, Payne SE, Haskins WCK (1988) J Am Chem Soc 110: 6240
159. Leis Chr, Lachmann G, Müller G, Zybill Chr, Polyhedron, in print
160. Hoa Tran Huy N, Mathey F (1987) Organometallics 6: 207
161. Chang LS, Corey JY (1989) Organometallics 8: 1885
162. a) Corey JY, Chang LS, Corey ER (1987) Organometallics 6: 1595
    b) Corey JY, Zhu XH, Bedard TC (1990) IXth International Symposium on Organosilicon Chemistry, 16–20 July, Edinburgh, Abstr. B. 10

162. c) Harrod JF, Ziegler T, Tschinke V (1990) Organometallics 9: 897
163. Campbell WH, Hilty TK (1989) Organometallics 8: 2615
164. Woo HG, Tilley TD (1989) J Am Chem Soc 111: 8043
165. a) Auburn M, Ciriano M, Howard JAK, Murray M, Pugh NJ, Spencer JL, Stone FGA, Woodward P (1980) J Chem Soc Dalton Trans 659
     b) Herrmann WA, Voss E, Guggolz E, Ziegler ML (1985) J. Organomet Chem 284: 47
166. Bennet MJ, Simpson KA (1971) J Am Chem Soc 93: 7156
167. Crozat MM, Watkins SF (1972) J Chem Soc Dalton Trans 2512
168. Cowie M, Bennett MJ (1977) Inorg Chem 16: 2321, 2325
169. Simon GL, Dahl LF (1973) J Am Chem Soc 95: 783
170. Hencken G, Weiss E (1973) Chem Ber 106: 1747
171. Schmid G, Welz E (1979) Z Naturforsch 34b: 929
172. Bondi A (1964) J Phys Chem 68: 441
173. Anderson AB, Shiller P, Zarate EA, Tessier-Youngs CA, Youngs WJ (1989) Organometallics 8: 2320
174. Grev RS, Schaefer III HF (1987) J Am Chem Soc 109: 6577
175. Powell J, Sawyer JF, Shiralian M (1989) Organometallics 8: 577
176. Tobita H, Kawano Y, Shimoi M, Ogino H (1987) Chem Lett 2247
177. Herrmann WA (1982) Adv Organomet Chem 20: 159
178. Casey CP, Audett JD (1986) Chem Rev 86: 339
179. Hahn JE (1984) Progr Inorg Chem 31: 205
180. Holten I, Lappert MF, Pearce R, Yarrow PIW (1983) Chem Rev 83: 135
181. Moss JR, Scott LG (1984) Coord Chem Rev 60: 171
182. Ozawa F, Park JW, Mackenzie PB, Schaefer WP, Henling LM, Grubbs RH (1989) J Am Chem Soc 111: 1319
183. Cowie M, McDonald R, Antonelli DM (1989) Interfaces in metal complex chemistry, International Symposium, Rennes, France, July 9–14, Abs. C 13
184. Bianconi PA, Weidman TW (1988) J Am Chem Soc 110: 2342
185. Jarrold MF, Bower JE (1989) J Am Chem Soc 111: 1979
186. Morancho R, Pouvreau P, Constant G, Jaud J, Galy J (1979) J Organomet Chem 166: 329
187. von Schnering HG (1984) Angew Chem 93: 44; (1981) Angew Chem Int Ed Engl 20: 33
188. Herzog S, Krebs F (1963) Naturwiss 8: 330
189. Huttner G, Weber U, Sigwarth B, Scheidsteger O, Lang H, Zsolnai L (1985) J Organomet Chem 282: 331
190. Matsumoto H, Higuchi K, Hoshino Y, Koike H, Naoi Y, Nagai Y (1988) J Chem Soc Chem Commun 1083
191. Kabe Y, Kuroda M, Yamashita O, Kawase T, Masamune S (1988) Angew Chem 100: 1793; (1988) Angew Chem Int Ed Engl 27: 1793
192. Sekiguchi A, Kabuto C, Sakurai H (1989) Angew Chem 101: 97; (1989) Angew Chem Int Ed Engl 28: 55
193. Weidenbruch M, Grimm FT, Pohl S, Saak W (1989) Angew Chem 101: 201; (1989) Angew Chem Int Ed Engl 28: 198
194. Bock H (1989) Angew Chem 101: 1695; (1989) Angew Chem Int Ed Engl 28: 1690
195. $Si(CO)_2$ is a loose donor adduct of CO and Si
196. Grev RS, Schaefer III HF (1989) J Am Chem Soc 111: 5687
197. Lembke RR, Ferrante RF, Weltner Jr W (1977) J Am Chem Soc 99: 416
198. ref. to: Beck W, Knauer W, Robl Chr (1990) Angew Chem 102: 331; (1990) Angew Chem Int Ed Engl 29: 318
199. Lei D, Hampden-Smith MJ, Duesler EN (1990) 199th ACS National Meeting, 22–27 April, Boston, INOR 448
200. Anema SG, Mackay KM, Nicholson BK (1989) J Organomet Chem 371: 233
201. Barrau J, Hamida NB, Agrebi A, Satge J (1989) Organometallics 8: 1585

The Use of π-Organoiron Sandwiches in Aromatic Chemistry

Didier Astruc

Laboratoire de Chimie et Organométallique, UA CNRS N° 35, Université de Bordeaux I, 351 cours de la Libération, 33405 Talence Cédex, FRANCE

Table of Contents

Topics in Current Chemistry, Vol. 160
© Springer-Verlag Berlin Heidelberg 1991

D. Astruc

The temporary complexation of aromatics by the $FeCp^{n+}$ or $FeC_6R_6^{m+}$ units (R = H or CH_3; n = 0 or 1; m = 0–2) activates the formation of many C–C and C-element bonds. This principle and its current applications are reviewed here. In the 18-electron sandwich complexes $Fe^{II}Cp(arene)^+$ and $Fe^{II}(arene)_2^{++}$, advantage has been taken of the electron-withdrawing properties of the cationic moiety for (i) deprotonation-alkylation (ii) nucleophilic addition and substitution. Nucleophilic displacement of the halide(s) in halogeno- or dihalogenoarene complexes $FeCp(arene)^+$ by amines, RO^-, SR^-, and stabilized carbanions was applied to the synthesis of heterocycles and of triaryldiethers. The multiple deprotona-tion-alkylation of $FeCp(C_6Me_6)^+$ gives tentacled aromatic complexes which can be hexafunctional. In the $Fe(arene)_2^{++}$ series, a few double alkylations have been performed, a strategy limited by the side electron-transfer pathway. Protection by hydride allows the reactions of a variety of carbanions. Deprotection is carried out using $Ph_3C^+BF_4^-$: electron-transfer gives the intermediate 17-electron radical cation and $Ph_3C^·$ followed by spontaneous H-atom transfer. In the non-methylated series, a second carbanion attack opens the route to heterobifunctional cyclohexadienes with a precise regio- and stereocontrol. In the crowded permethylated series, a second functionalization is achieved by deprotonation-acylation. In each series, the polyene ligand is disengaged using oxidation of the functional sandwich with $Al_2O_3 + O_2$. Activation at the 19-electron or 20-electron stage proceeds by electron-transfer to the substrate and subsequent cage reaction. The isoelectronic 19-electron complexes $Fe^ICp(arene)$ and $Fe^I(cyclohexadienyl)(arene)$ react in this way with O_2 with leads to benzylic activation, a reaction inhibited by the salt effect. Subsequent functionaliza-tion can be achieved with many elements. In the 20-electron complex $Fe^0(C_6Me_6)_2$, double C–H activation by O_2 gives an *ortho*-xylylene complex and electron-transfer to functional halides gives functional cyclohexadienyl complexes. The long known redox reactions of $FeCp(arene)^+$ complexes at the benzylic position of the arene substituent are also briefly reviewed. The low cost of the activating iron moieties, the ease of complexation and decomplexation, the non-toxicity, and the possibility of using the temporary complexation for multiple activation steps make the iron-mediated reactions a powerful tool in synthetic strategies.

1 Introduction

The discovery of ferrocene [1–3] prompted the development of organometallic chemistry [4] and subsequent applications to organic chemistry [5]. Yet the iron atom in ferrocene did not provide much organic synthesis of iron-free cyclopentenoid compounds although the latter would have been of interest as an entry to the prostaglandin families. Ferrocene can be considered as a tridimentional superatomatic [6] with its own, well-developed, organic chemistry [7]. On the other hand, the molecular transformations achieved via temporary complexation to a transition-metal group are not only extensively developed in catalysis but also in stoichiometric activation [8]. The advantage of the latter is that the transition metal, held in the organometallic framework, can be used in several consecutive steps. Therefore, two concepts illustrating the properties of transition metals are applied. The first one is the activation of organic molecules by π-coordinated electron-withdrawing transition − metal groups. This complexation can reverse the property of the organic molecule. For example, aromatic compounds are usually electrophilic but their coordination to $Cr(CO)_3$ [9], $FeCp^+$ [10–13] or $Mn(CO)_3^+$ [13, 14] inhibits their electrophilic properties and brings about nucleophilic properties [13]. It is possible to rank the different transition-metal activators in 18-electron complexes according [14] to their electron-withdrawing properties characterized by the relative rates of their reactions with nucleophiles (Fig. 1).

The second concept is to switch the reactivity of the complex by addition or substraction of one or two electrons to or from the complex; applies the "Umpolung" principle. The effect of redox change is dramatic: increase in the rate is often of the order of 10^9 for a given reaction [15–18]. These properties are summarized in the following Scheme I:

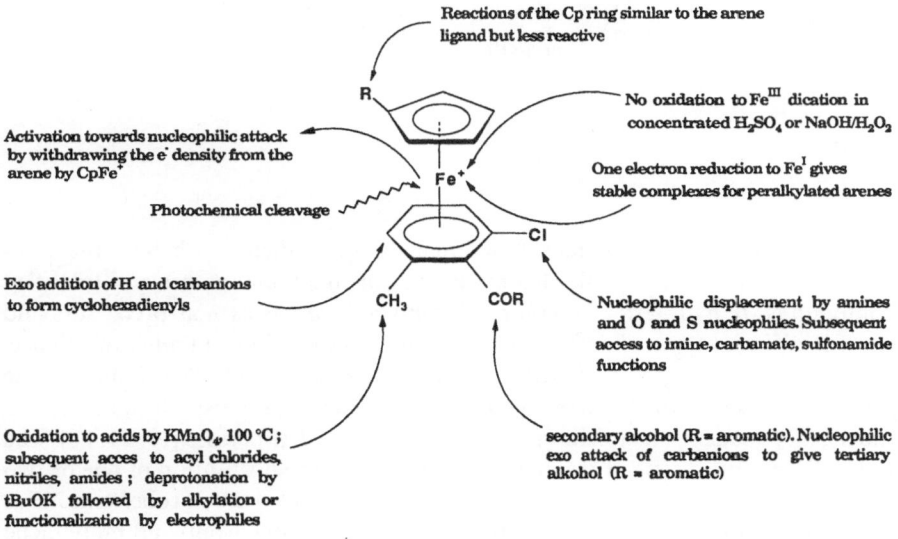

Reactions of the Cp ring similar to the arene ligand but less reactive

No oxidation to Fe^{III} dication in concentrated H_2SO_4 or $NaOH/H_2O_2$

Activation towards nucleophilic attack by withdrawing the e⁻ density from the arene by CpFe⁺

One electron reduction to Fe^I gives stable complexes for peralkylated arenes

Photochemical cleavage

Exo addition of H⁻ and carbanions to form cyclohexadienyls

Nucleophilic displacement by amines and O and S nucleophiles. Subsequent access to imine, carbamate, sulfonamide functions

Oxidation to acids by $KMnO_4$, 100 °C; subsequent acces to acyl chlorides, nitriles, amides; deprotonation by tBuOK followed by alkylation or functionalization by electrophiles

secondary alcohol (R = aromatic). Nucleophilic exo attack of carbanions to give tertiary alcohol (R = aromatic)

Scheme I

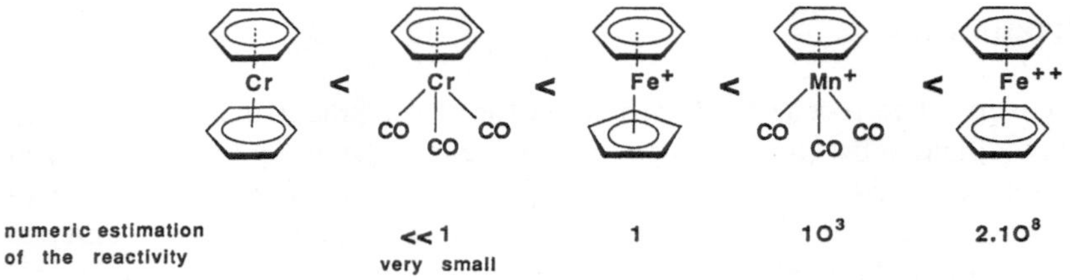

Fig. 1.

The activation of aromatics by organo-iron groups has been used with two families of complexes [10–13], both synthesized from the arene and $AlCl_3$ with variation in the source of iron only [19–20] as shown in Scheme II:

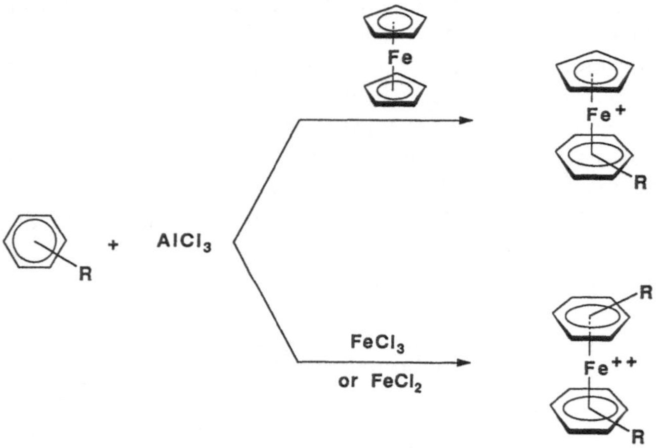

Scheme II

The symmetric series provides functional cyclohexadienes, whereas the non-symmetric one serves to build deuterated and/or functional arenes and tentacled compounds. *In both series, several oxidation states can be used* as precursors and provide different types of activation. The complexes bearing a number of valence, electrons over 18 react primarily by electron-transfer (ET). The ability of the sandwich structure to stabilize several oxidation states [21] also allows us to use them as ET reagents in stoichiometric and catalytic ET processes [18, 21, 22]. The last well-developed type of reactions is the nucleophilic substitution of one or two chlorine atoms in the $FeCp^+$ complexes of mono- and o-dichlorobenzene. This chemistry is at least as rich as with the $Cr(CO)_3$ activating group and more facile since $FeCp^+$ activator is stronger than $Cr(CO)_3^-$.

For former reviews on parts of the topic, see Refs. [10–12]. The subject is increasingly attractive and has now far-reaching applications in organic synthesis and supramolecular chemistry.

2 Bis-arene Iron Di-cations: Nucleophilic Attacks

Bis-arene iron dications [20] are easily accessible from arenes, $AlCl_3$ and $FeCl_3$ (for C_6Me_6, $FeCl_2$ must be used). It is advisable to use tris-sublimated $AlCl_3$ to avoid problems of isomerization [23]. With toluene, this isomerization due to the *retro*-Friedel-Crafts mechanism [24] is too extensive to give any clean complex.

The parent bis-benzene complex is hydrolytically sensitive [20] and must be used in its crude form. All the reactions with carbanions give ET products. First, the purple 19e mono-cation $[Fe(C_6Me_6)_2]^+$ [25–28] is formed immediately at $-90\,°C$. Then the black 20e complex $Fe(C_6Me_6)_2$ is observed [25–29]. However, recently, Zaworotko et al. have succeeded in making a C–C bond using $AlEt_3$ as the carbanion source [30].

The analogue $[Fe(C_6H_6)(C_6Me_6)]^{++}(PF_6^-)_2$ [23] was made by double hybride abstraction from the known cyclohexadiene complex $[Fe(\eta^4\text{-}C_6H_8)(\eta^6\text{-}C_6Me_6)]$ [29]. It behaved in the same way, although the reactions were smoother.

Prior to our work, Helling had reacted a series of hard carbanions with $[Fe(mesitylene)_2]^{++}(PF_6^-)_2$ to obtain an attack on each ring giving the bis(cyclohexadienyl)Fe^{II} complex [31], Eq. (1). Stabilized carbanions reacted only once [31, 32]; first step of Eq. (1):

$$\text{(1)}$$

Davies et al. used these results to try to show the validity of the charge control rule in the nucleophilic attack on the mixed cations (bearing both an odd and an even ligand) [33, 34]. We suspected that this orientation was indeed due to the special location in the 1, 3, 5 positions of the methyl groups around the arene ligand and decided to investigate the reactivity of nucleophiles on the parent cation $[Fe^{II}(\eta^5\text{-cyclohexadienyl})(\eta^6\text{-}C_6H_6)]^+$. We found that a large variety of nucleophiles cleanly react on this complex to give exclusively substituted cyclohexadiene Fe^0 complexes in high yields [35]. Moreover we also demonstrated, using deuterium labelling experiments, that the thermodynamic product is also the kinetic one [36]. Finally, extended Hückel calculations by Saillard and Hoffmann showed that orbital control is in agreement with the observed regiospecificity, as

D. Astruc

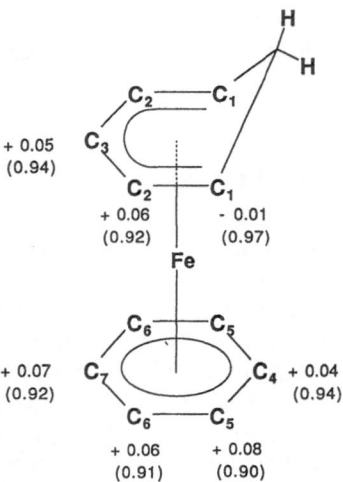

+ 0.05
(0.94)

C$_2$ — C$_1$
C$_3$

C$_2$ — C$_1$
+ 0.06   − 0.01
(0.92)   (0.97)

Fe

C$_6$ — C$_5$
C$_7$ C$_4$ + 0.04
+ 0.07   (0.94)
(0.92)

C$_6$ — C$_5$
+ 0.06   + 0.08
(0.91)   (0.90)

Fig. 2. Total charges and, in parentheses, P$_\pi$ populations of carbon atoms in $[Fe(C_6H_6)(\eta^5\text{-}C_6H_7)]^+$

opposed to the explanation involving charge control [36]. Thus the charge was calculated on each ligand carbon, confirming that the C$_1$ positions adjacent to the sp^3 carbons bear less charge than the other ligand carbon (Fig. 2). The molecular orbital diagram of $[Fe(\eta^6\text{-}C_6H_6)(\eta^5\text{-}C_6H_7)]^+$ is shown below (Fig. 3).

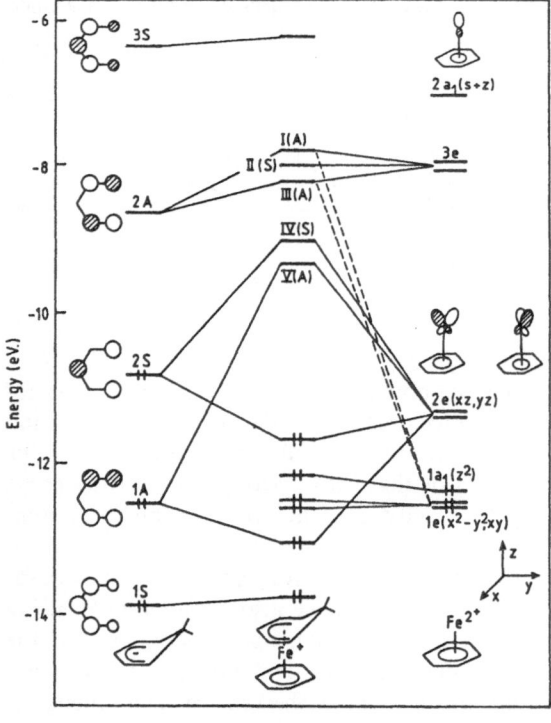

Fig. 3. Molecular interaction diagram for $[Fe(C_6H_6)(\eta^5\text{-}C_6H_7)]^+$

Table 1. Relative energies and H ... C overlap populations for the system of H· approaching $[Fe(C_6H_6)(\eta^5\text{-}C_6H_7)]^+$ at various (H ... C = 2.0 Å)

	H_{ii} for $H_{(ev)}^-$	Total energy of the system (relative value with respect to H^- attack at C_1)	H^- ... C overlap population
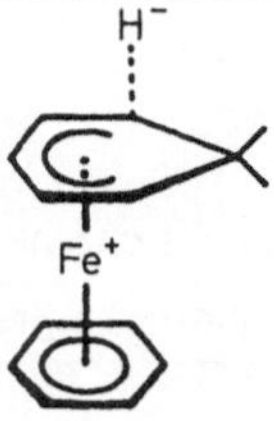 (structure: H^-, $\overset{..}{C}$, Fe^+, ring)	-13.6	0.00	0.034
	-11.0	0.00	0.137
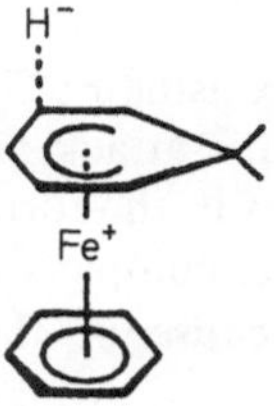 (structure: H^-, $\overset{..}{C}$, Fe^+, ring)	-13.6	$+0.11$	0.001
	-11.0	$+0.31$	0.080
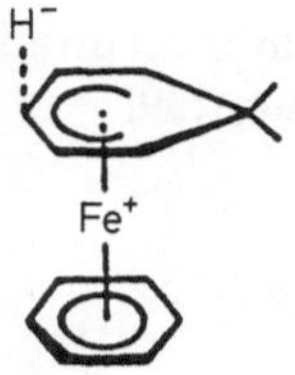 (structure: H^-, $\overset{..}{C}$, Fe^+, ring)	-13.6	$+0.05$	0.017
	-11.0	$+0.17$	0.102
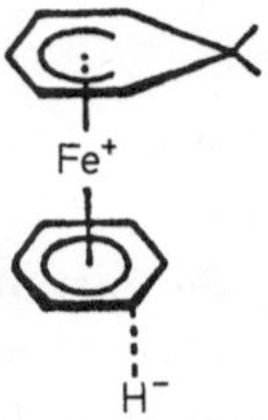 (structure: $\overset{..}{C}$, Fe^+, ring, H^-)	-13.6	$+0.06$	0.021
	-11.0	$+0.20$	0.102

The following Table 1 shows the relative energies and H ... C overlap populations for the system of H^- approching the cationic complex at various positions.

On the other hand, $NaBH_4$ reacts cleanly with $[Fe(C_6H_6)_2]^{++}PF_6^-{}_2$ in THF at 0 °C to give the hydride transfer product $[Fe(\eta^5\text{-}C_6H_7)(\eta^6\text{-}C_6H_6)]^+$ [37] and another equivalent of the hydride can be used to make the neutral cyclohexadiene complex $[Fe(\eta^4\text{-}C_6H_8)(\eta^6\text{-}C_6Me_6)]$ [35, 36]. The monocation also reacts cleanly with a variety of carbanions under the form of lithium, sodium and potassium or Grignard reagents in good yields to give monosubstituted cyclohexadiene iron complexes [35, 36], Eq. (2).

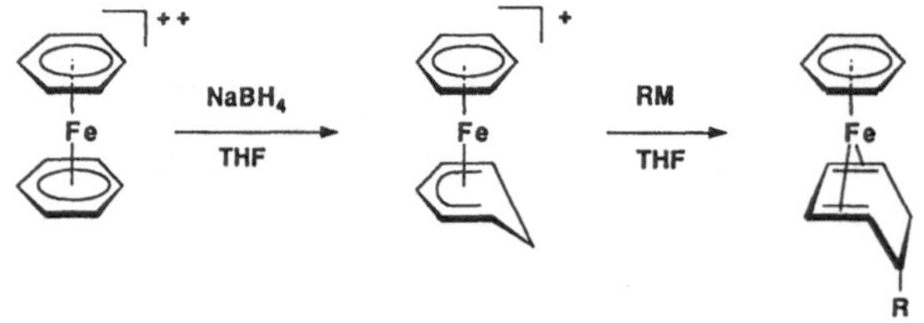

$$\text{(2)}$$

RM	NaBH$_4$	KCN	NaCH(CO$_2$Et)$_2$	PhCH$_2$MgBr	LiCHS(CH$_2$)$_3$S	Cp(η^5-CH$_2$C$_6$Me$_5$)Fe
R	H	CN	CH(CO$_2$Et)$_2$	CH$_2$PH	CHS(CH$_2$)$_3$S	Cp(η^5-CH$_2$C$_6$Me$_5$)Fe
Yield %	90	85	75	55	65	59

The organic ligands can be obtained by oxidation of the iron complex using FeCl$_3$ in ether [35, 36]. This chemistry parallels the well-known nucleophilic attack on cyclohexadienyl iron tricarbonyl cations which needs soft carbanions in the form of zinc- or cadmium reagents [38]. Hydride abstraction from the latter complexes, to regenerate a functional cationic complex, does not work because of the cumulative bulk of R$^-$ and the reagent Ph$_3$C$^+$ [38].

We similarly attempted to remove the hydride in the sandwich series and room-temperature experiments faced the same steric problem. Reactions occurred by an ET pathway followed by cleavage of the C–C bond in the 17e cation [39].

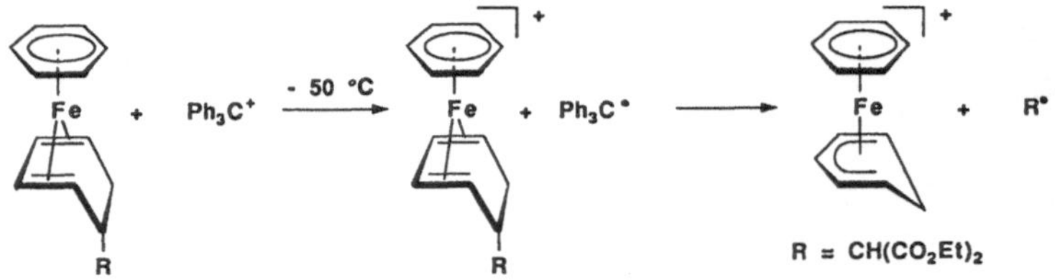

$$\text{(3)}$$

$$R = CH(CO_2Et)_2$$

However, when the reaction was effected at $-40\,°C$ and the reaction mixture kept for 3 hours at $-40\,°C$, the ET produced a 17e cation which did not decompose but was characterized by ESR and smoothly transferred a hydrogen atom to Ph$_3$C$^{\cdot}$. The reaction between the two organic and organometallic radicals (observed by ESR) is much easier than hydride transfer between closed-shell compounds and does not suffer from the steric restriction of the latter [39].

Thus, functional cyclohexadienyl complexes were obtained in this way which allowed a second reaction with a nucleophile. For instance, with KCN in acetone, the yield was 87% (Fig. 4). Bifunctionalization could indeed be performed to synthesize *exo,exo*-1,2-disubstituted cyclohexadiene Fe^0 complexes. These regio- and stereocontrols were confirmed by the x-ray crystal structure of the cyano-benzyl complex (Fig. 4). Decomplexation of the Fe^0 complex using Al$_2$O$_3$ + O$_2$ gave the free ligand characterized by its mass spectrum

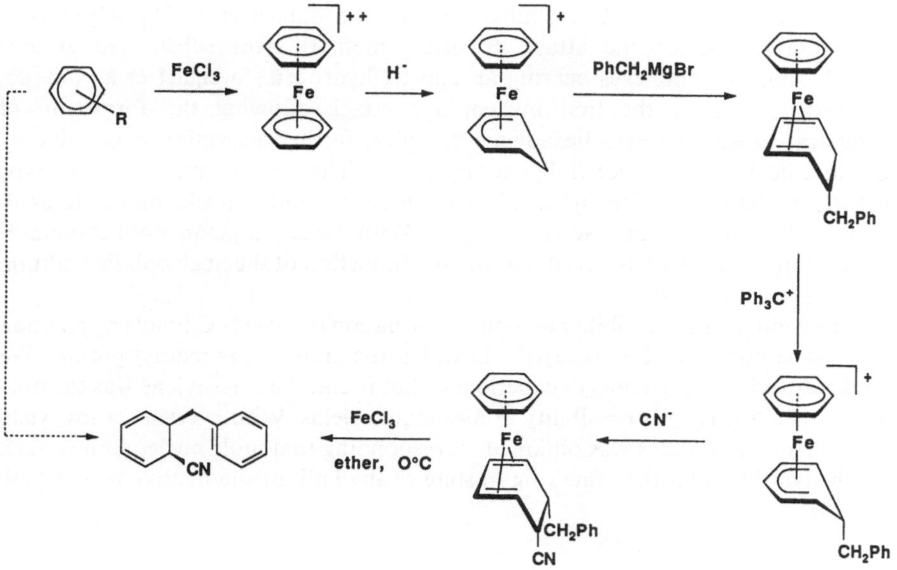

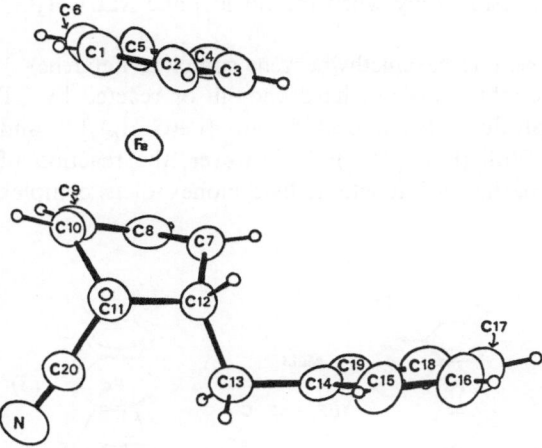

Scheme III: Hydride reservoir: protection by hydride for the transformation of benzene into hetero-bifunctional cyclohexadienes

(aromatization occurred at 40 °C in the mass spectrum showing the bifunctional arene) [39]. The overall strategy is summarized in Scheme III.

Thus, heterobifunctional cyclohexadienes are accessible from benzene by temporary complexation to Fe^{2+} which reverses the reactivity of the aromatic. Double nucleophilic attack is possible but only using protection-deprotection by hydride.

Fig. 4. Ortep view of $[Fe(\eta^6\text{-}C_6H_6)(\eta^4\text{-}exo, exo\text{-}5\text{-}(PhCH_2)\text{-}6\text{-}CNC_6H_6)]$

The $Mn(CO)_3^+$ unit allows a similar strategy. Brookhart et al. [40–42] showed that double nucleophilic attack (hydride, methyl) gives substituted anionic cyclohexadiene complexes bearing an agostic hydrogen. Sweigart et al. [43–45] showed that, after the first nucleophilic attack allowing the formation of a neutral substituted cyclohexadienyl complex, further activation is possible by displacement of a carbonyl ligand by NO^+. The new complex of the type $[Mn(CO)_2(NO)(\eta^5\text{-}RC_6H_6)]^+$ can be submitted to another nucleophilic attack to give a substituted cyclohexadiene complex. With the chiral piano stool complexes $[Mn(CO)(PPh_3)(NO^+)(C_6H_7)]^+$, asymmetric induction of the nucleophilic addition was demonstrated [45].

We examined the possibility of a direct formation of two C–C bonds by reaction of a carbanion with $[Fe(arene)_2]^{2+}$ in which the arene bears methyl groups. We could indeed repeat Hellings's experiments but found that mesitylene was the only aromatic allowing this possibility in reasonable yields. With p-xylene, a low yield of an unstable complex was obtained corresponding to double nucleophilic attack of phenyllithium on the same ring in spite of the bulk of the methyl groups [23]. Eq. (4):

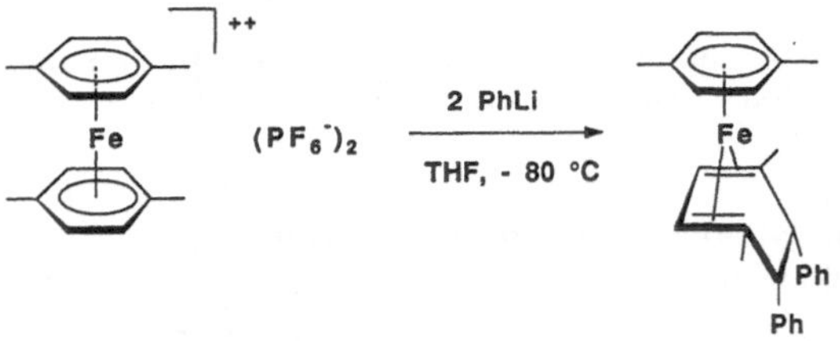

$$(4)$$

Such double C–C bond formation had also been attempted by Vollhardt et al. by nucleophilic attack on $[CoCp(C_6H_6)]^{++}$. Formation of disubstituted cyclohexadiene complexes was possible only when the nucleophile was CH_3O^- or $C_5H_5^-$.

The durene, pentamethylbenzene, and hexamethylbenzene complexe $Fe(arene)_2^{++}$ either did not react if the nucleophile was not hard enough or reacted by ET to give the 19e and the 20e products (as indicated with $[Fe(C_6H_6)_2]^{2+}$ and $[Fe(C_6Me_6)(C_6H_6)]^{2+}$ above). With $[Fe(C_6Me_6)_2]^{2+}$, however, the reaction of excess methyllithium in THF gave the neutral octamethylcyclohexadiene complex in low yield, Eq. (5):

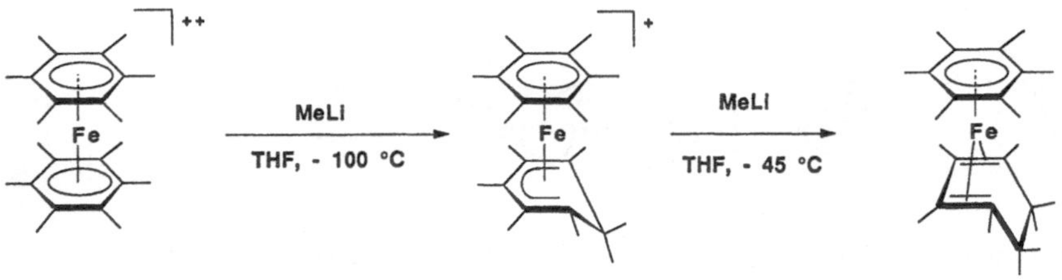

$$(5)$$

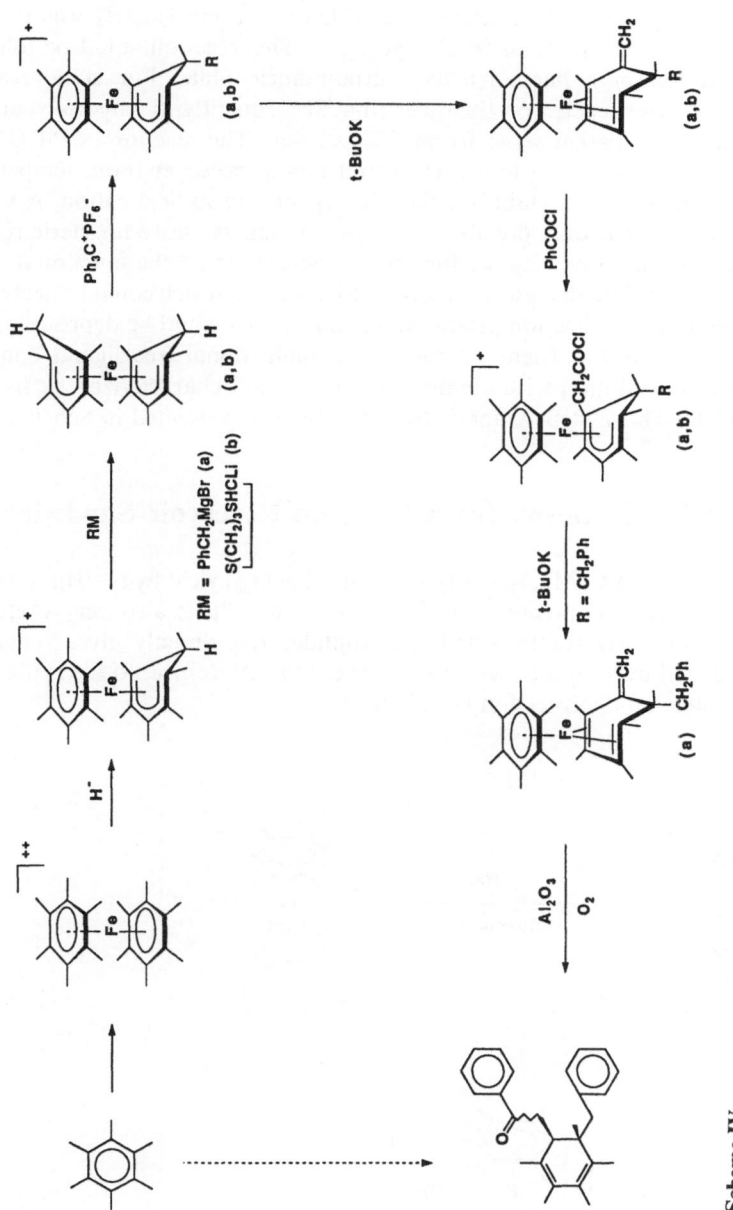

Scheme IV

57

Thus as in the parent complex, addition of hydride using $NaBH_4$ was the only way to enter the chemistry of $[Fe(C_6Me_6)_2]^{2+}$. This reaction could be followed by the addition of a hard, reactive carbon nucleophile. For steric reasons, however, the latter reacted on the other ring. Subsequently, the hydride could be removed as in the parent series by an ET pathway. The reaction of $Ph_3C^+BF_4^-$ to remove H^- sequentially $(e^- + H^{\cdot})$ could now proceed at room temperature since the permethylation stabilizes the $17e$ organoiron radical cation. A second nucleophilic attack is only possible on the other ring as above for steric reasons and was not further investigated. Instead, we deprotonated the functional cation to obtain an arene ligand with an exocyclic double bond which could be acetylated. This second functionalization produced a cation which could be deprotonated to give a birefringent Fe^0 triene complex. The bifunctional free ligand could be disengaged by oxidation with alumina under O_2, and characterized by its mass spectrum [46]. This 8-step synthesis from C_6Me_6 is represented in Scheme IV.

3 Use of ET Pathway from Electron-Reservoir Sandwiches

The $20e$ complex $Fe(C_6Me_6)_2$, easily synthesized in high yield by Na/Hg reduction of the dry di-cationic precursor in THF at 20 °C [28], is also very useful for functionalization. Its reaction with electrophiles RX directly gives functional cyclohexadienyl iron cations, which saves one step with respect to the route using hydride protection/deprotection [47]; Scheme V:

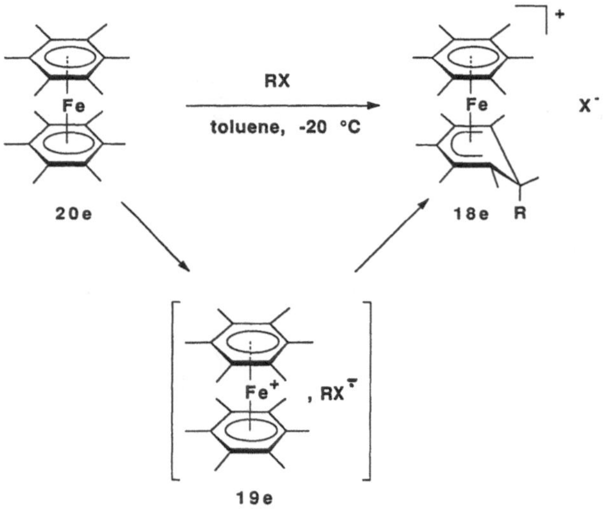

R (Yield %) : CH_2Ph (90), COPh (90), CH_2CN (40), CH_2CO_2Me (50), $CH_2CH=CH_2$ (50)

Scheme V

The reaction proceeds by an ET pathway giving the 19e organoiron radical cation and the organic radical R˙ which couple in the cage after escape of X⁻. The cationic Fe^I intermediate is noted at low temperature by its characteristic purple color and the classical spectrum of Fe^I species with rhombic distortion (g = 2.091, 2.012, 2.003 at −140 °C in acetone) before collapse to the orange substituted cyclohexadienyl Fe^{II} complexes.

The 19e electron-reservoir complexes Fe^{I}Cp(arene) can give an electron to a large number of substrates and several such cases have been used for activation. After ET, the [Fe^{II}Cp(arene)]$^+$ cation left has 18 valence electrons and thus cannot react in a radical-type way in the cage as was the case for 20e Fe^0(arene)$_2$ species. Thus the 19e Fe^{I}Cp(arene) complexes react with the organic halide RX to give the coupled product and the [FeCp(arene)]$^+$ cation. Only half of the starting complex is used; e.g., the theoretical yield is limited to 50% [48] (Scheme VI) contrary to the reaction with Fe^0(arene)$_2$ above.

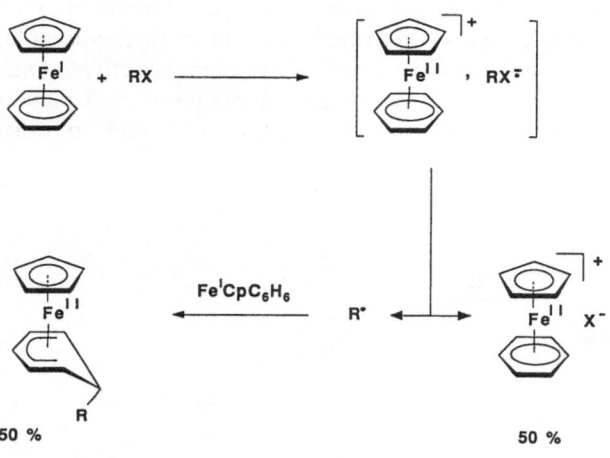

Scheme VI

However, if the radical anionic substrate produced in such a type of process reacts with the cationic partner within the cage in a non-radical way, the reaction is no longer marred by the above difficulty and subsequent yield limitation.

The most spectacular activation obtained in this way is the one using molecular oxygen: ET to O$_2$ produces O$_2^{\bar{}}$ which reacts in the cage as a base or as a nucleophile [21]. The primary ET from Fe^I to O$_2$ is extremely fast because it is exergonic. The E^0 value of Fe^I/Fe^{II} is around −1.5 V vs SCE on Hg cathode in DMF and that of O$_2$/O$_2^{\bar{}}$ is around −0.8 V vs SCE under these conditions.

When the iron sandwich complex bears an arene substituent with at least one benzylic hydrogen, the acidity of the latter is enhanced by the π-complexation to the 12e fragment FeCp$^+$. The pK_a of the conjugate acid of superoxide radical anion, $HO_2^·$ is only 4.8 [49], but nevertheless O_2^- behaves as a base because it can deprotonate weak acids with up to pK_a 22 [50]. This is due to the large dismutation ability of $HO_2^·$ which displaces the acido-basic equilibrium. Thus, H^+ as well as transition metal ions (for example in the superoxide dismutase enzymes) can catalyse the disproportionation of O_2^- [51–53]. The reaction with Fe^I complexes is a in Eq. (6):

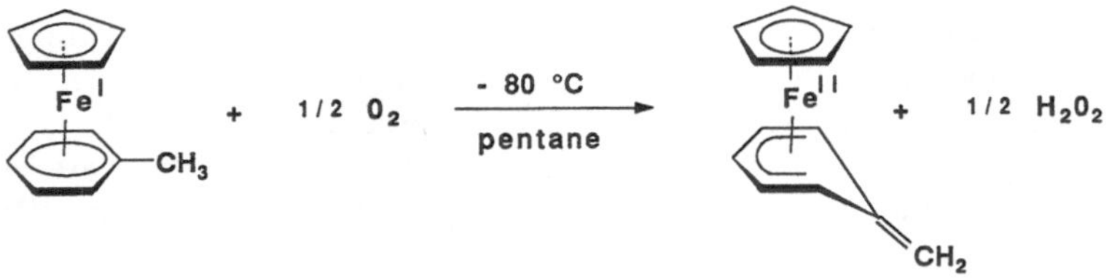

$$\tag{6}$$

Subsequent functionalization is detailed in the following section. The reaction of the Fe^I complex with O_2 was effected with various arene ligand structures. With the C_6R_6 ligand (R = Me or Et), the C–H activated products are thermally stable [54, 55]. With $C_6Me_5NH_2$ ligand, the more acidic N–H bond is activated and the resulting imino complex can react with CO_2 to give an organo-iron carbamate zwitterion [56], a reaction which could be extended to CS_2 and to other electrophiles Eq. (7)

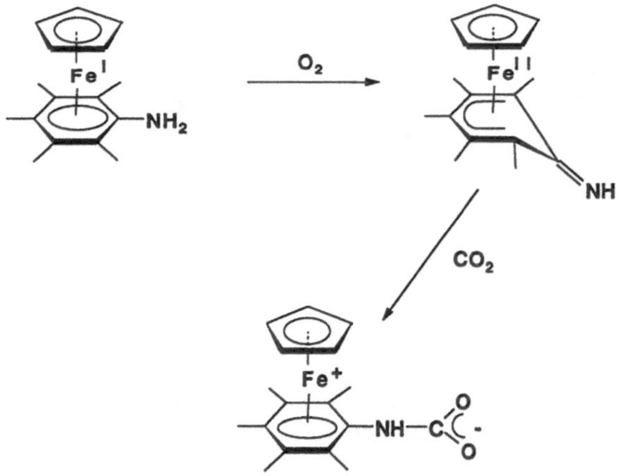

$$\tag{7}$$

Similarly, when both the Cp* and arene ligands are permethylated, the reaction of O_2 with the Fe^I complex leads to C–H activation of the more acidic benzyl bond [57]. When no benzylic hydrogen is present, superoxide reacts as a nucleophile and adds onto the benzene ligand of the $FeCp(arene)^+$ cation to give a peroxocyclohexadienyl radical which couples with a $Fe^ICp(arene)$ radical. A symmetrical bridging peroxo complex $[(Fe^{II}Cp)_2(\eta^5-C_6H_6O)_2]$ is obtained. The C–H activation reactions of the $19e$ Fe^I radicals BH can be summarized as follows

[58]; Eq. (8):

$$BH^{\cdot} + O_2 \rightarrow [BH^+, O_2^{\overline{\cdot}}] \rightarrow B + HO_2^{\cdot}$$
$$\downarrow \qquad\qquad (8)$$
$$O_2 + H_2O$$

or if the Fe^I complex A does not contain a benzylic hydrogen (Eq. (9)):

$$A^{\cdot} + O_2 \rightarrow [A^+, O_2^{\overline{\cdot}}] \rightarrow AOO^{\cdot}$$
$$\downarrow A^{\cdot} \qquad\qquad (9)$$
$$AOOA$$

Both the basic and nucleophilic reactions within the cage are totally inhibited by the presence of one equivalent of $Na^+PF_6^-$ in THF. Double ion exchange between the two ion pairs is favored which removes superoxide from the cage; Eq. (10):

$$Fe^I Cp(arene) + O_2 \rightarrow [Fe^{II}Cp(arene)^+, O_2^{\overline{\cdot}}]$$
$$\Big| \quad Na^+, PF_6^- \qquad\qquad (10)$$
$$[Fe^{II}Cp(arene), PF_6^-] + Na^+, O_2^{\overline{\cdot}}]$$

Sodium peroxide [2 Na^+, $O_2^{\overline{}}$] is obtained. This indicates that Na^+ also induces the disproportionation of superoxide [59], presumably due to the large difference of interaction energy between the contact ion pair $[Na^+, O_2^{\overline{\cdot}}]$ and the solvent-separated ion pair $[Na^+/THF/O_2^{\overline{\cdot}}]$.

The special salt effect is a constant feature of the activation of substrates in cages subsequent to ET from electron-reservoir complexes. In the present case, the salt effect inhibits the C–H activation process [59], but in other cases, the result of the special effect can be favorable. For instance, when the reduction of a substrate is expected, one wishes to avoid the cage reaction with the sandwich. An example is the reduction of alkynes and of aldehydes or ketones [60]. These reductions follow a pathway which is comparable to the one observed in the reaction with O_2. In the absence of $Na^+PF_6^-$, coupling of the substrate with the sandwich is observed. Thus one equiv. $Na^+PF_6^-$ is used to avoid this cage coupling and, in the presence of ethanol as a proton donor, hydrogenation is obtained (Scheme VII).

$$Fe^I Cp(C_6Me_6) + S \rightarrow [Fe^{II}Cp(C_6Me_6)^+, S^-]$$
$$[Fe(C_6Me_6)\eta^4\text{-}Cp - S] \longleftarrow \Big|$$
$$\Big\downarrow [Na^+, PF_6^-], C_2H_5OH$$
$$[Fe^{II}Cp(C_6Me_6)]^+PF_6^- + SH_2$$

Scheme VII   S = PhC≡CH, PhCHO, Me$_2$CO

Thus, the knowledge of the salt effect gained from the above mechanistic studies is useful to apply C–H activation by O_2 to other systems under suitable conditions. The reaction of the 20e complex $Fe^0(C_6Me_6)_2$ with O_2 in toluene, in the absence of sodium salt, involves double H atom abstraction giving an o-xylylene complex; Eq. (11). The mechanism is also an ET from the 20e complex to O_2 giving deprotonation of an acidic methyl group of the cation by O_2^-. The neutral 19e intermediate again reacts similarly with O_2 as does the Fe^I analog $[Fe^I(\eta^5\text{-}C_6Me_6H)(\eta^6\text{-}C_6Me_6)]$; Eq. (12). Electron-transfer from Fe^0 to O_2 is fast because of the large difference of E^0 values between the two reversible redox system (1.4 V vs SCE res. 0.8 V vs SCE in DMF). Subsequent deprotonation by O_2^- within the cage has a Fe^I intermediate, the structure of which is very close to that of the Fe^I complex in Eq. (12). The E^0 value for the latter is -1.45 V vs SCE on Hg cathode in DMF. Thus, electron-transfer from these Fe^I species to O_2 is also exergonic, thus fast.

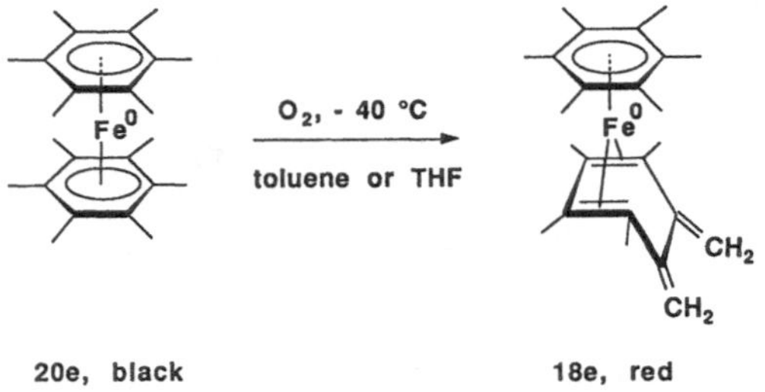

$$\text{(11)}$$

20e, black   18e, red

$$\text{(12)}$$

19e brown   18e red

Although the o-xylylene complex is thermally unstable, it was characterized at $-50\ °C$ by its ^1H- and ^{13}C-NMR spectra showing the exocyclic methylene at $\delta = 5.04, 4.42$ ppm (^1H) and $\delta = 144.8$ ppm (^{13}C) using $C_6D_5CD_3$ as the solvent. Its reaction with benzoyl chloride on the exocyclic carbon leaves a very acidic methylene group which transfers a proton onto the adjacent methylene unit. The double bond is benzoylated again in **in situ** and a di-cation of the $[bis(arene)Fe]^{2+}$ type is obtained [47]; Scheme VIII.

In all the working strategies with the $[bis(arene)Fe]^{2+}$ series, only one ring must be functionalized several times, although the starting product is symmetric.

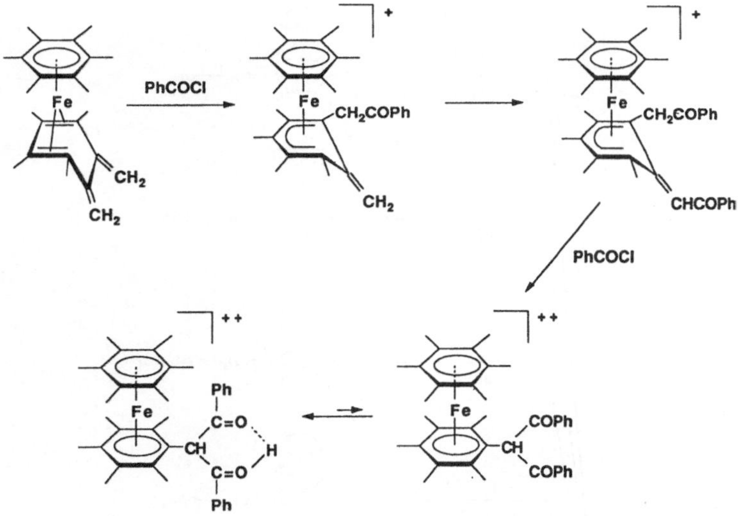

Scheme VIII

4 CpFe(arene)$^+$ Mono-cations: Benzylic Deprotonation

The acidity of benzylic protons of aromatics complexed to transition-metal groups was first disclosed by Trakanosky and Card with (indane)Cr(CO)$_3$ [61]. Other cases are known with Cr(CO)$_3$ [62], Mn(CO)$_3^+$ [63], FeCp$^+$ [64, 65], and Fe(arene)$^{2+}$ [31, 66] but none reported the isolation of deprotonated methyl-substituted complexes. We found that deprotonation of the toluene complex gives an unstable red complex which could be characterized by ^{13}C NMR ($\delta_{CH_2} = 4.86$ ppm vs TMS in CD$_5$CD$_3$) and alkylated by CH$_3$I [58]; Eq. (13):

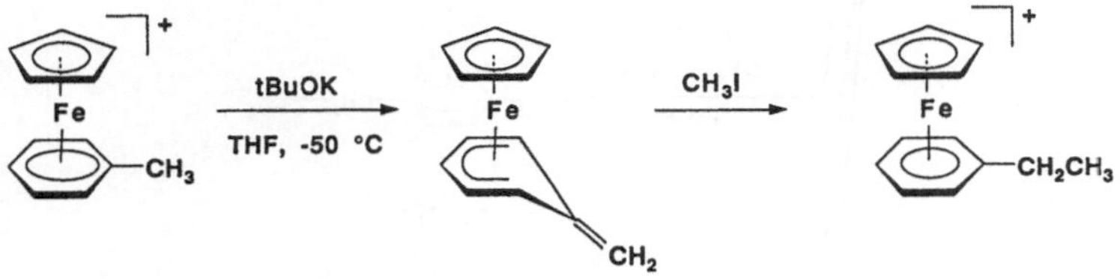

(13)

The hexamethylbenzene complex could be similarly deprotonated [54, 57] but the red complex obtained is then stable and its x-ray crystal structure could be recorded [55], showing a dihedral angle of 32° between the cyclohexadienyl plane and the *exo*cyclic double bond (Fig. 5). This complex can also be cleanly obtained by the reaction of dioxygen with the 19*e iso*structural complex Fe^{I}Cp(C$_6$Me$_6$) as shown in the preceding section.

D. Astruc

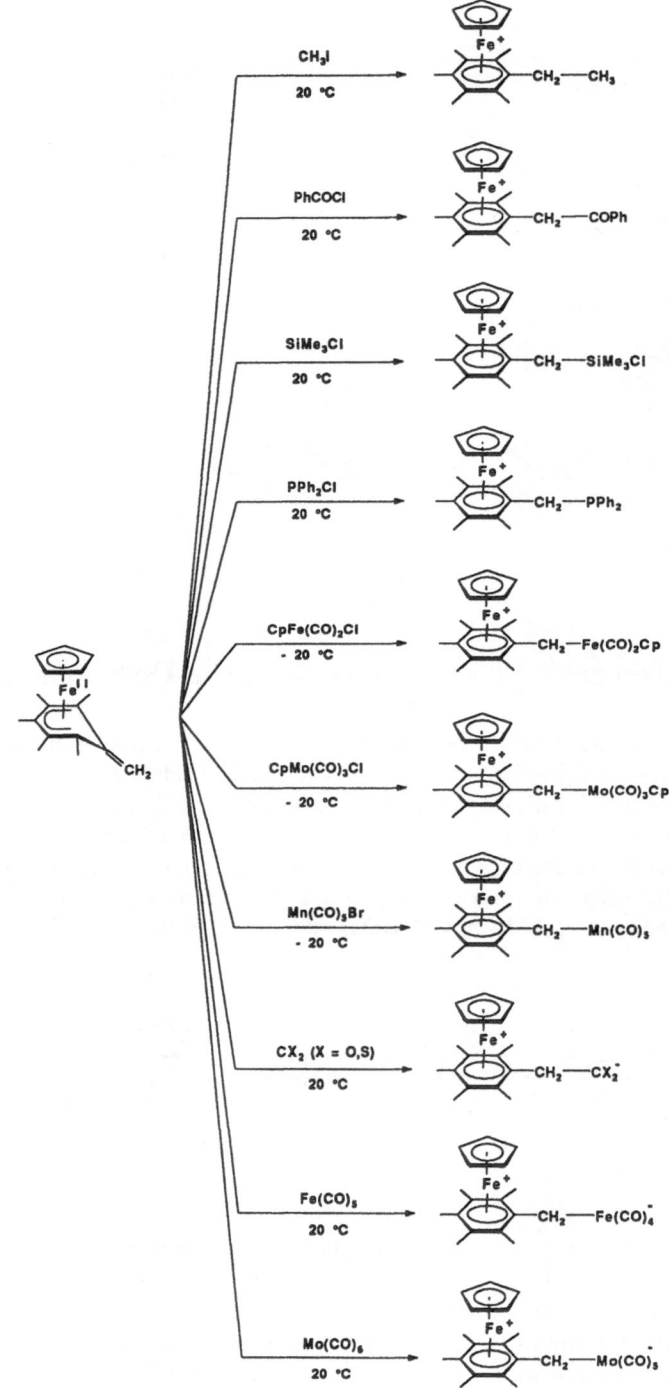

Scheme IX

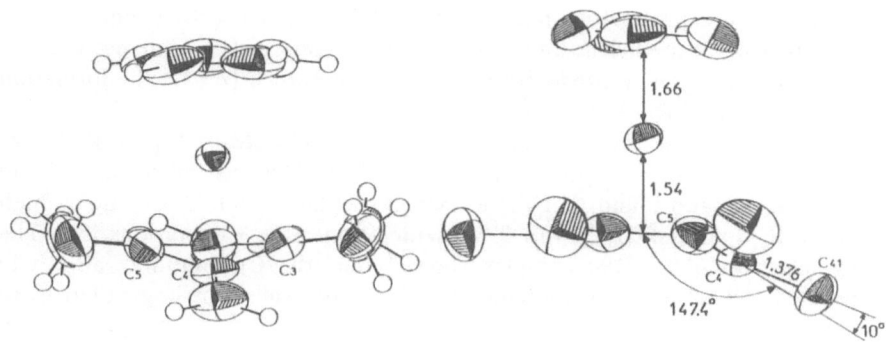

Fig. 5. Front and side view of the complex $Fe^{II}CpC_6Me_5(CH_2)$ resulting from C–H abstraction with O_2

The deprotonated complex is a nucleophilic synthon which reacts smoothly with a large variety of organic and inorganic electrophiles forming a carbon-element bond; Eq. (14) and Scheme IX:

$$[FeCp(\eta^5\text{-}C_6Me_5CH_2)] + RX \rightarrow [FeCp(\eta^6\text{-}C_6Me_5CH_2R)]^+X^- \quad (14)$$

Before deprotonation of the unstabilized alkylarene complexes, deprotonation of FeCp(arene)$^+$ cation with a conjugated benzylic position was observed. The first case was reported by Treichel using FeCp(fluorene)$^+$ and t-BuOK [67, 68]. In the blue zwitterion obtained, the angle of 11° between the planes of the complexed − and uncomplexed rings was recorded by x-ray. Contrary to the Mn(CO)$_3$ complex, no shift of complexation to the central C$_5$ ring of fluorene was observed. Methylation was very slow with MeI (1 day; 20 °C) but faster using CH$_3$SO$_3$F (a few minutes; 20 °C) giving the *exo*-methyl fluorene complex. Alternatively, Fe^{I}Cp(fluorene) reacted with O$_2$ to give H atom abstraction at low temperature (-80 °C) [69]. At 20 °C, H-atom abstraction can even occur spontaneously in the absence of O$_2$ [70]. The *endo*-H-atom abstraction by O$_2$ works equally well in the *exo*-9-methyl derivative which allowed a second methylation. The reaction with PhCOCl does not give the *exo*-9-benzoyl derivative because the very acid benzylic proton therein protonated the starting zwitterionic complex to give the bis-benzoylated derivative [69]. Deprotonation of the cationic complexes of tri-phenylmethane, diphenylamine, and carbazole was performed as well [71–73]. Then deprotonation and subsequent methylation of the deprotonated complex using MeI was reported for the complexes of aniline, phenol, and thiophenol to give the complexes of dimethylaniline, anisole, and methyl-phenyl sulfide.

The deprotonation of benzylic carbon facilitated by the Cr(CO)$_3$ has also been studied [18] by other authors after the pioneering demonstration of Trahanowsky and Card using this unit. Thus, it has been applied to the alkylation of phenylacetate, acetophenone, and ethylbenzene [19]. Similarly, bis(mesitylene)-

Fe^{++} has been deprotonated, but the reaction is complicated by further nucleophilic attack of the methylene unit with the starting material [17]. Enhanced acidity of the ring hydrogens in arene-metal complexes is shown [21] by the formation of complexes of alkyllithium by proton abstraction.

The stability of the deprotonated complex $FeCp(\eta^5\text{-}C_6Me_6CH_2)$ at 20 °C was promising in terms of multiple activation. We investigated the deuteration in D_2O + NaOD and found that the exchange was fast at 80 °C, leading to high yields of $[CpFeC_6(CD_3)_6]^+$ with 99% deuteration after 3 operations [74]. The mass spectrum of the 19e complex showed that the Cp ring was also fully deuterated. UV photolysis of the deuterated 18e cation in acetonitrile gave ferrocene d^{10} and $C_6Me_6d^{18}$ (Scheme X).

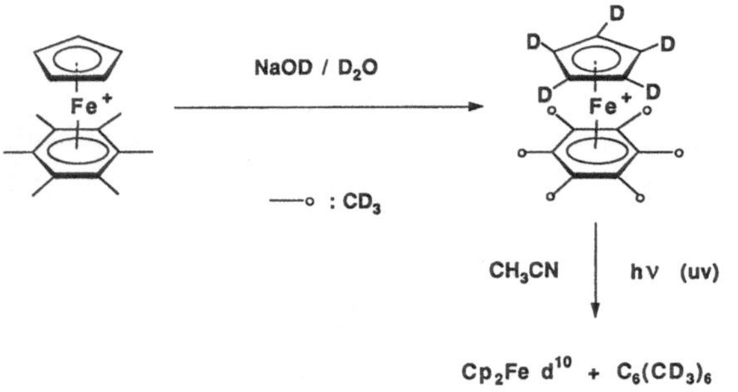

$$Cp_2Fe\ d^{10}\ +\ C_6(CD_3)_6 \qquad \textbf{Scheme X}$$

We wanted to know whether it was possible to effect the exchange of protons not only with deuterium but also with alkyl groups provided by alkyl halides. In other terms, we wished to extend the various functionalizations performed on a single methyl substituent to the six methyl groups of $[FeCp(C_6Me_6)]^+$.

The first attempted reaction was performed using t-BuOK and CH_3I, both reagents being in excess in THF. After a mild reflux of about 1 min, a quantitative yield of $[FeCp(C_6Et_6)]^+PF_6^-$ was obtained [75, 76]; Eq. (15). The reaction with CD_3I gave $[FeCpC_6(CH_2CD_3)_6]^+$ and the reactions of $[FeCpC_6(CD_3)_6]^+$ with CH_3I resp. CD_3I gave $[FeCpC_6(CD_2CH_3)_6]^+$ resp. $[FeCp(CD_2CD_3)_6]^+$.

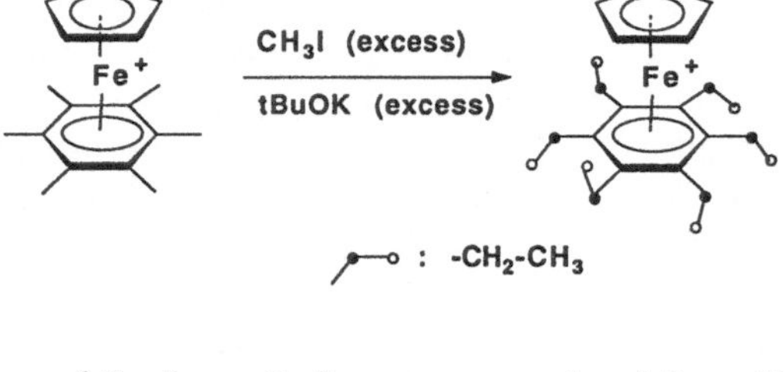

$$(15)$$

Deprotonation of the hexaethylbenzene complex *followed* by reaction with CH_3I in THF occurred only at reflux indicating that the 7th methylation is difficult but possible (Eq. (16)). Thus, t-BuOK and CH_3I react together more slowly than the first six alkylations of $[FeCp(C_6Me_6)]^+$ but more rapidly than the 7th one.

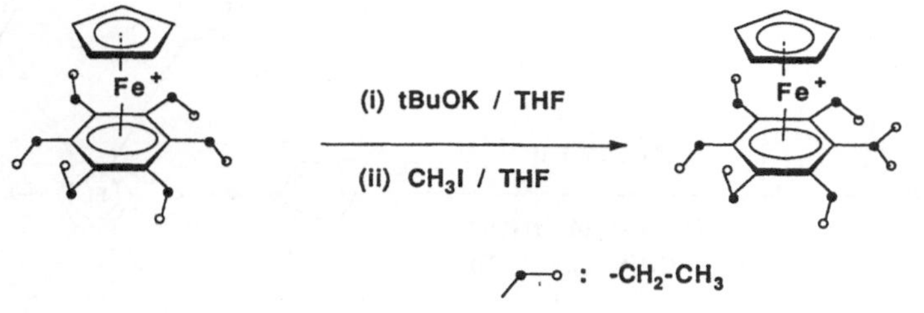

$$\text{(i) tBuOK / THF}$$
$$\text{(ii) } CH_3I \text{ / THF}$$

(16)

$$\nearrow\!\!\!-\!\!\!\circ \; : \; \text{-}CH_2\text{-}CH_3$$

The reason for the low rate of the 7th alkylation turns out to be steric since $[FeCp(mesitylene)]^+$ reacts with excess t-BuOK and CH_3I to give the pure 1,3,5-tris-t-butylbenzene complex directly; Eq. (17). The intermediate case is that of the durene complex which, under these reaction conditions, gives the pure 1,2,4,5-tetra-isopropylbenzene complex; Eq. (18). The o-xylene complex behaves in the same way; Eq. (19):

FE = C_5H_5Fe

(17)

FE = C_5H_5Fe

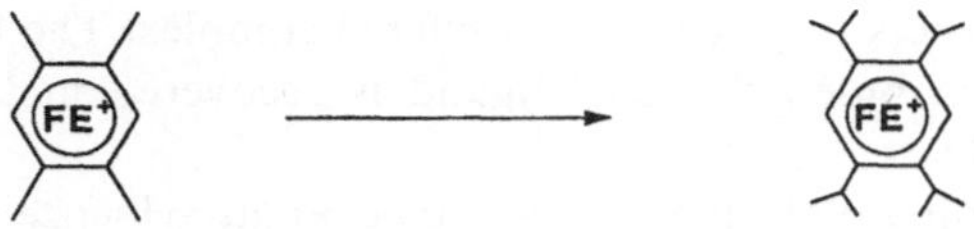

(18)

FE = C_5H_5Fe

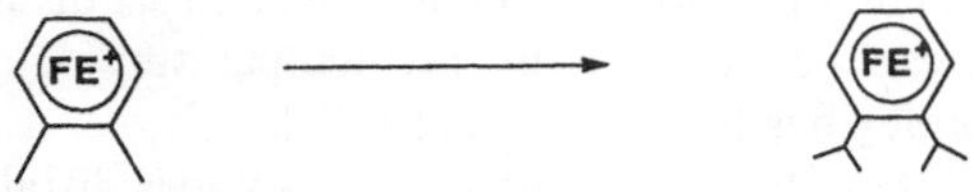

(19)

The reaction was extended to $PhCH_2X$ (X = Cl or Br) which gives the hexa(phenylethyl)benzene complex. The new free aromatic ligand is easily disengaged by photolysis in acetonitrile [76a]; Eq. (20). This line of research is now offering us the perspective of making new discotic liquid crystals using suitably substituted

67

D. Astruc

aryl derivatives [76b].

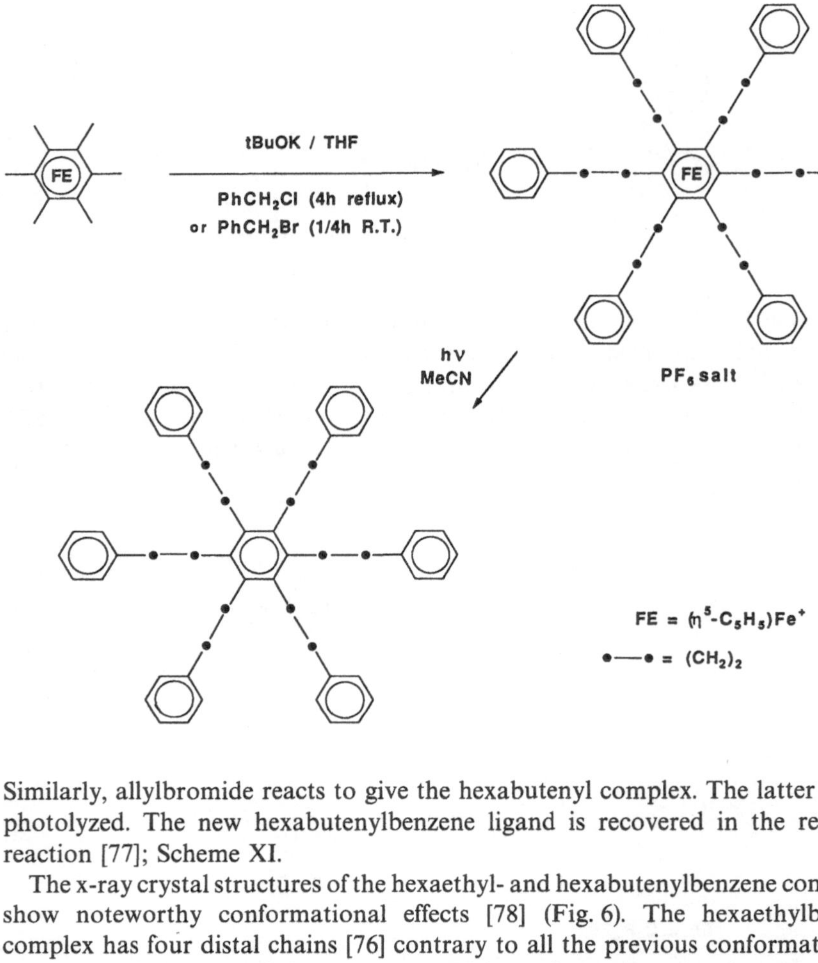

Similarly, allylbromide reacts to give the hexabutenyl complex. The latter can be photolyzed. The new hexabutenylbenzene ligand is recovered in the recycling reaction [77]; Scheme XI.

The x-ray crystal structures of the hexaethyl- and hexabutenylbenzene complexes show noteworthy conformational effects [78] (Fig. 6). The hexaethylbenzene complex has four distal chains [76] contrary to all the previous conformations of C_6Et_6 and $(M)C_6Et_6$ of C_{3v} symmetry. This conformation also depends on the counter-anion as the three conformations with four, five, and six distal ET groups have close energies and can be observed by low-temperature ^1H NMR. The hexabutenyl benzene complex has five distal chains [77].

Indeed, with the BPh_4^- counter-anion, there are also five distal ET groups in $FeCp(C_5Et_6)^+BPh_4^-$ [79]. On the other hand, the five Et groups of $FeCp^*(C_5Et_5H)^+PF_6^-$ are all distal (the latter complex was specifically obtained from $FeCp^*(CO)_2Br$ and C_6Et_6, which may open the route to penta-alkylbenzenes from hexa-alkyl benzenes [80]).

The hexabutenylbenzene complex reacts with Br_2 to give the dodecabromo-derivative, with bulky silanes and boranes to give regiospecific hydrosilylation and hydroboration (Scheme XII).

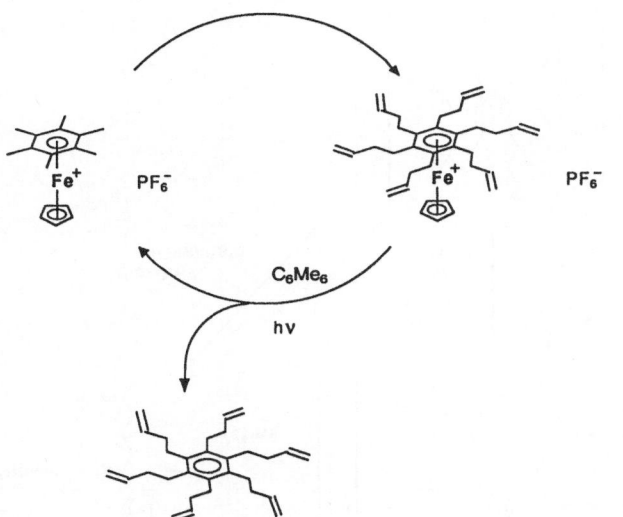

Scheme XI

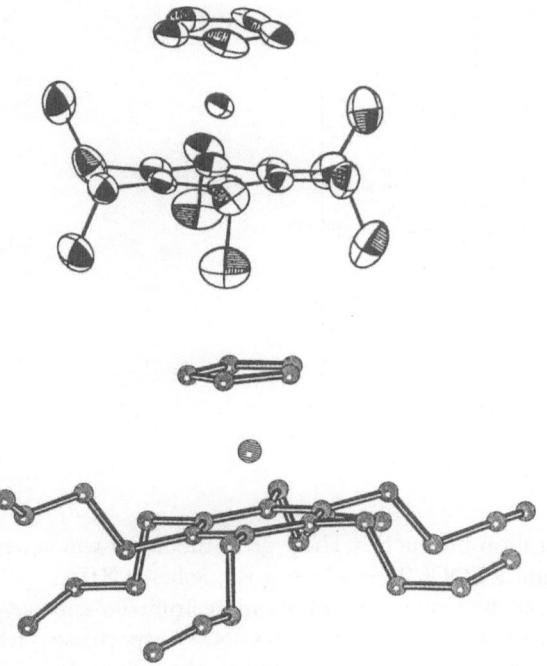

Fig. 6. Ortep views of [FeCp(hexaethylbenzene)]$^+$PF$_6^-$ and [FeCp(hexabutenylbenzene)]$^+$PF$_6^-$

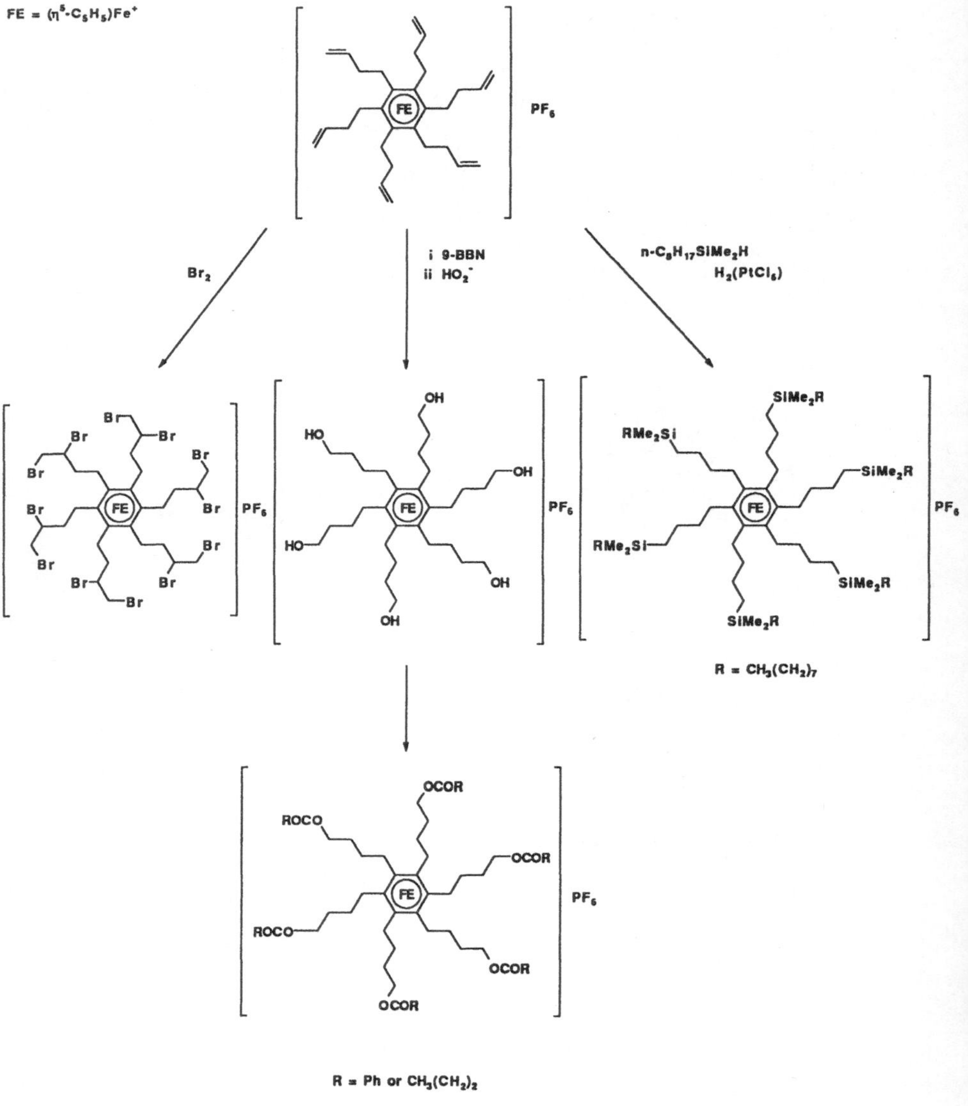

D. Astruc

FE = $(\eta^5\text{-}C_5H_5)Fe^+$

Scheme XII Syntheses of tentacled iron sandwiches

Oxidation of the hexahydroboration product by HO_2^- gives the hexol which can be esterified in neat acyl chloride RCOCl (R = Ph or *n*-Bu, Scheme XII).

The free ligands were obtained by visible photolysis in acetonitrile and were characterized by elemental analyses and ^{1}H- and ^{13}C-NMR spectroscopies. Low-temperature reduction in ESR tubes also provided the green Fe^I 19e complexes which gave the characteristic 3g values typical of the rhombic distortion of Fe^I [77].

A major limit of the hexaalkylation system with t-BuOK was the failure of the reactions with alkyl halides bearing at least one β hydrogen. Dehydrohalogenation by t-BuOK gave the olefin, the reaction (Eq. (21)) being always faster than the organometallic alkylation. Thus, $[[FeCp(C_6Me_6)]^+PF_6^-$ remained unchanged:

$$RCH_2-CH_2X + t\text{-BuOK} \rightarrow RCH_2=CH_2 + t\text{-BuOH} + K^+X^-$$

$$(R = H, X = I).\tag{21}$$

This difficulty could be overcome by using KOH instead of t-BuOK. No reaction between KOH and RX was noted and, although the organometallic reaction is somewhat slower, it proceeds smoothly. Thus, this technique opens a general route to the complexes of hexaalkylbenzene with a variable chain length. So far, the hexapropyl, hexabutyl, and hexahexylbenzene complexes were made in a pure form. The hexaalkylation reaction tolerates an oxygen atom in the chain (ether group) provided it is remote from the alkylation site [81a]; Scheme XIII. Efforts

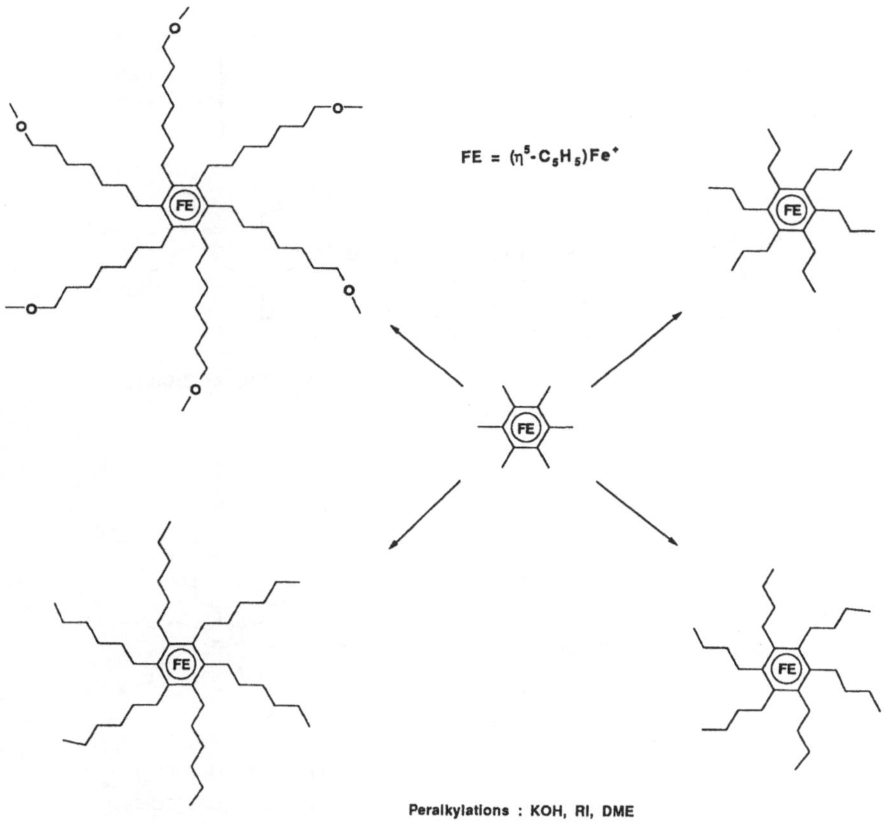

FE = $(\eta^5\text{-}C_5H_5)Fe^+$

Peralkylations : KOH, RI, DME

Scheme XIII

are continuing along these lines to make polyether derivatives, polyfunctional complexes, hexaligands, and templates [81 b].

Both the hexaolefin and its FeCp$^+$ complex were hydrozirconated using ZrCp$_2$(H)(Cl) which gave regiospecifically the hexazirconated compounds. The latter are air-sensitive and are easily hydrolyzed to hexabutylbenzene and its FeCp$^+$ complex. Both the hexaborane [78] and the hexazirconated [82] derivatives reacted with I$_2$ at 20 °C to give the hexaiodo derivatives but the latter was only obtained pure using the hexaborane (Scheme XIV).

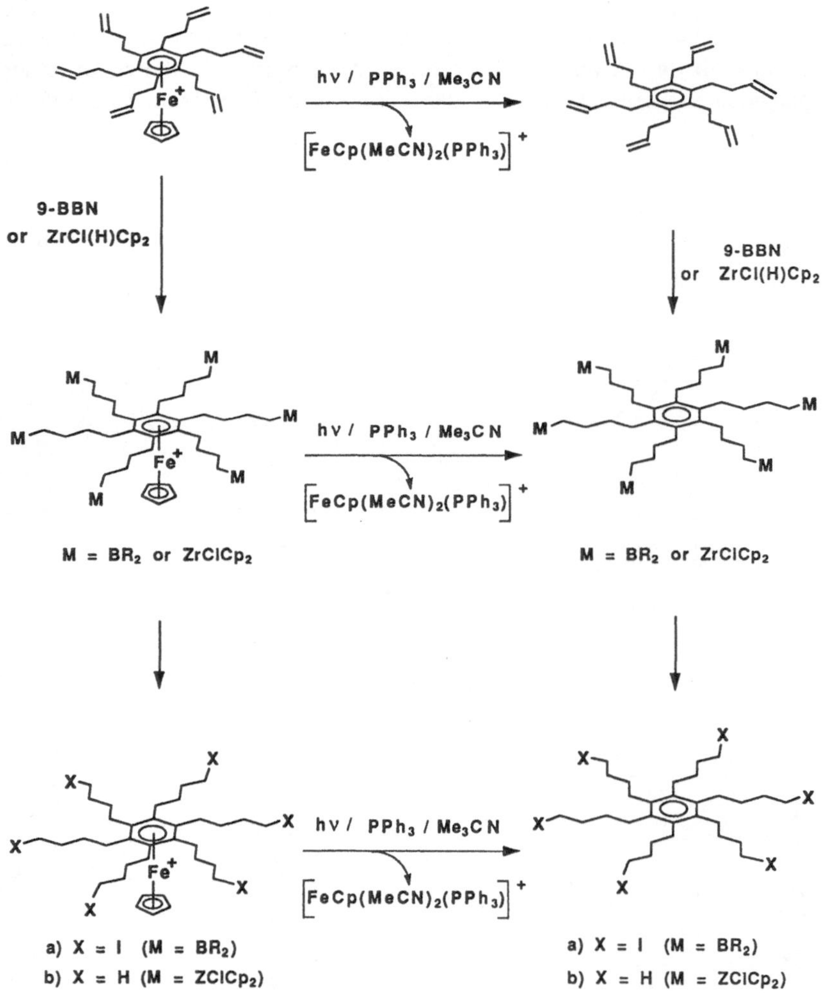

Scheme XIV

5 Activation of Halogenoaromatics by FeCp$^+$ in the SN$_{Ar}$ Reaction

5.1 Monochloroaromatics

More than twenty years ago, Nesmeyanov's group showed that chlorine can be substituted by a variety of nucleophiles in $FeCp(\eta^6\text{-PhCl})^+$ [83, 84]. Indeed the chlorine substituent in the chlorobenzene (even) ligand is 1000 times more reactive than when it is located on the cyclopentadienyl (odd) ligand [85]. The FeCp$^+$ is a good withdrawing group which is equivalent to two nitro groups in terms of activation. The reactions proceed under ambient conditions with primary or secondary amines and have been extended to other substituted chloroarene complexes [86, 87]; Eq. (22), Table 2.

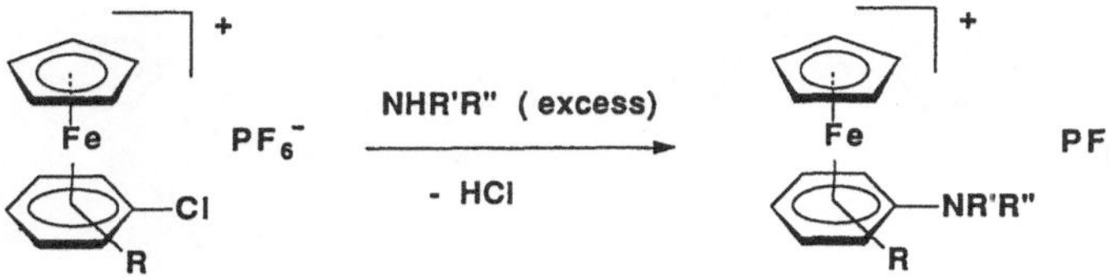

$$ \text{(22)} $$

Table 2. Yield of nucleophilic substitution of chlorine in $[FeCp(C_6H_5Cl)]^+$ by amino groups using amines at 20 °C

R	NR'R"				
	MeNH-	NH$_2$CH$_2$CH$_2$NH-	C$_4$H$_8$N-	C$_6$H$_{11}$-NH-	C$_6$H$_5$CH$_2$NH-
H	81%	70%	88.5%	68%	70%
o-CH$_3$	86%	65%	85%	48%	80%
m-CH$_3$	85.5%	68–70%	77%	75%	80%
p-CH$_3$	85%	81.8%	77%	72.3%	77.5%

Another remarkable reaction is the nucleophilic substitution of the chlorine by alkoxy or sulfido groups using the alcohol or the thiol and the weak base Na$_2$CO$_3$ in situ. For example, in the case of ethanol, the reaction proceeds in 12 h at reflux; Eq. (23), Table 3.

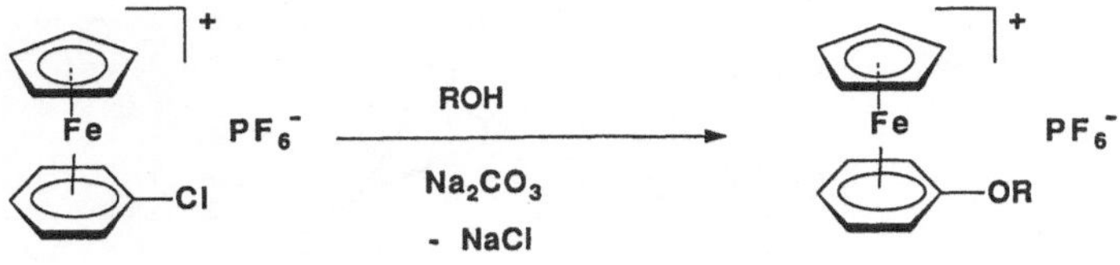

$$ \text{(23)} $$

D. Astruc

With OH^- and SH^-, the nucleophilic substitution of Cl has been reported. Thus, with NaOH, there is a report of successful nucleophilic substitution in 50% aq. acetone at room temperature to give the phenol complex in 36% yield. The latter is then spontaneously deprotonated to give the cyclohexadienyl complex (Eq. (24)). An identical reaction was carried out using NaSH in MeCN (50% yield) to give the thiophenol complex which was deprotonated [72]; Eq. (25). These reactions would be especially valuable because direct synthesis of the phenol or thiophenol complexes from ferrocene is not possible due to the strong interaction between the heteroatom and $AlCl_3$ [11, 19]. Recent improvement and use of this reaction were achieved [88].

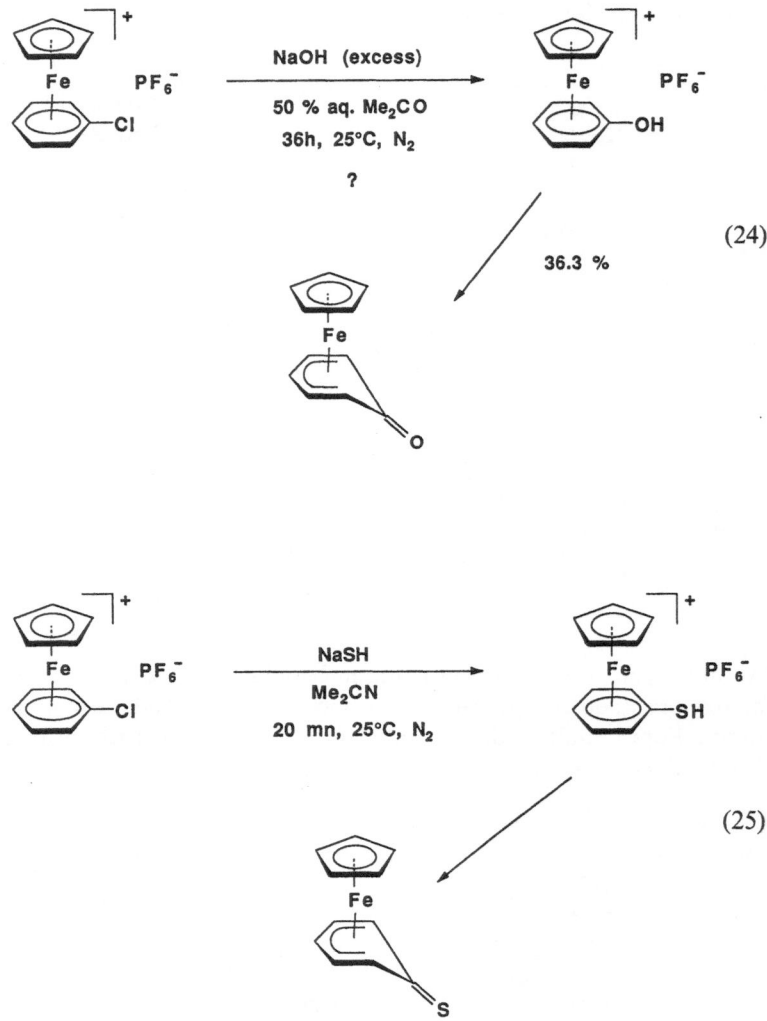

Table 3

solvent	EtOH	Me$_2$CO	EtOH	EtOH	Me$_2$CO
ROH or RSH	EtOH	PhOH	PhSH	BuS	Phthalimido
Yield %	72	82	88	89	?

Similarly, Lee et al. have substituted the chlorine by azoturo in several chloroarene complexes [89, 90]; Eq. (26), Table 4.

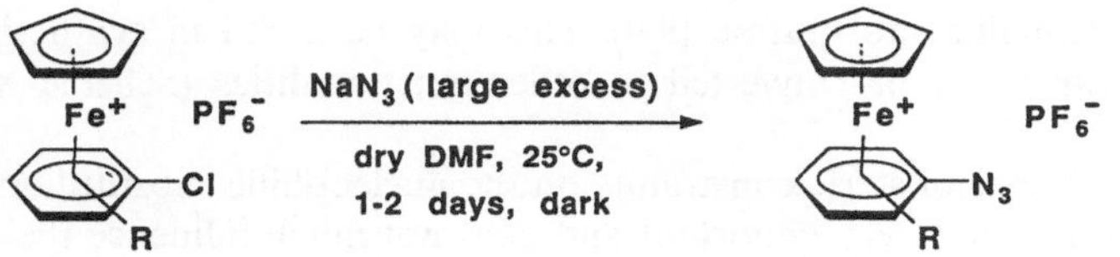

$$(26)$$

Table 4. Yield of chlorine substitution by azoturo in chloroarene complexes

R	H	o-CH$_3$	m-CH$_3$	p-CH$_3$
Rdt %	83	80	80	86

A kinetic study of the nucleophilic substitution of Cl$^-$ in [FeCp(PhCl)]$^+$BF$_4^-$ by these different N-, O-, and S-anionic nucleophiles was performed using the titration of displaced Cl$^-$ by AgNO$_3$. The reactivity order increases according to the sequence:

$$NH_3^- \ll PhO^- < PhS^- < MeS^- < MeO^-$$

which can be correlated neither with the polarizability of the nucleophile nor with the relative reactivities with 2,4-dinitrobenzene. For neutral nucleophiles, the order follows:

$$thiourea < aniline < morpholine < piperidine < guanidine$$

The difference of reactivity between 2,4-dinitrochlorobenzene and the iron complexes is explained by the fact that the additional electronic density is localized on the oxygen atoms in β-exocyclic position in the case of the nitro derivatives and on the ring in the case of the iron complexes [91].

Recently, Moriary and Gill have reported the nucleophilic substitution of Cl^- by stabilized carbanions, thus allowing the formation of C–C bond formation between an arene and an aliphatic residue [92, 93]. The carbanions used are stabilized in the benzylic positions by two carbonyl groups and are generated by deprotonation of diethylmalonate, 2,4-pentanedione, dimedone, ethyl diacetatoacetate, benzylacetone, phenylsulfonylacetone, phenylsulfonylethyl acetate, and deoxybenzoin. The suggested mechanism involves reversible *ipso*-addition on the aromatic carbon atom bearing the leaving group Cl^- followed by irreversible loss of HCl giving the cyclohexadienyl iron complex which bears an *exo*cyclic double bond. It should be noted that only the stabilized carbanions do lead to nucleophilic substitution, whereas stabilized carbanions (Me^- in MeLi) exclusively added to the benzene ligand in the *ortho*-position to give he cyclohexadienyl derivatives [94, 95]. Similarly, carbanions stabilized by a nitro group in α position can also give nucleophilic substitution [95]). This may be useful in synthesis since nitro groups can easily be converted to other functionalities (Scheme XV, Tables 5 and 6).

The influence of steric constraints on the nucleophilic substitution of Cl^- by a nucleophile is not very important and does not much influence the yield even in the presence of two *ortho*-methyl groups. Steric inhibition was only found when

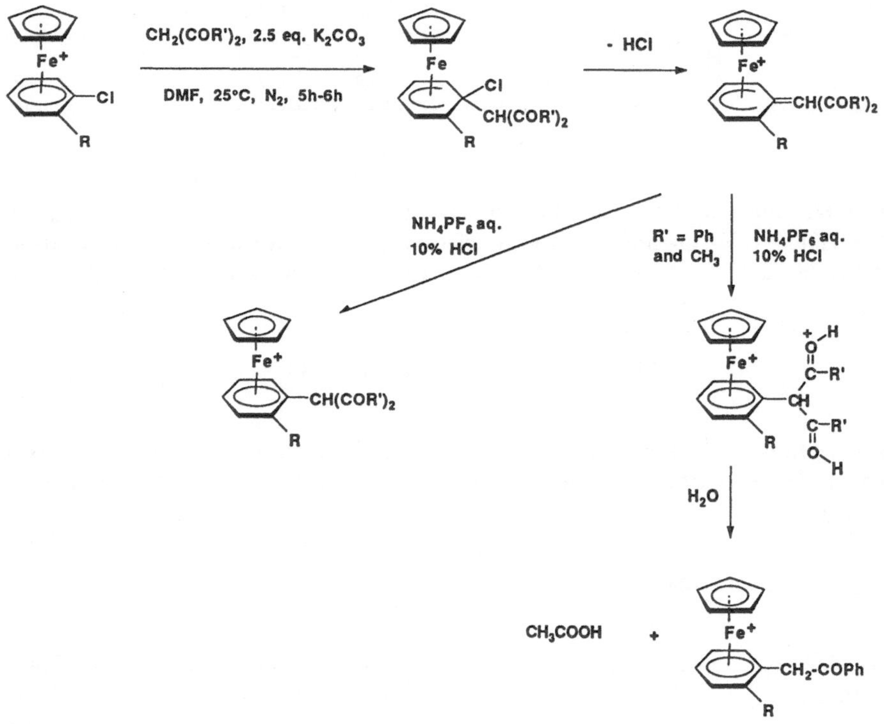

Scheme XV

Table 5

$(COR')_2$	Yield %
$(COCH_3)_2$	50
$(COCH_3)(COOC_2H_5)$	80
$(SO_2C_6H_5)(COCH_3)$	50
$(SO_2C_6H_5)(COOC_2H_5)$	80
$(COC_6H_5)_2$	65
$(COOC_2H_5)_2$	77
$(CH_2(COR)_2$ = dimedone	83

Table 6

R	Yield %
H	65
p-CH_3	65

the nucleophile was too large, as in the case of dibenzoylmethane [96]; Eq. (27) and Table 7.

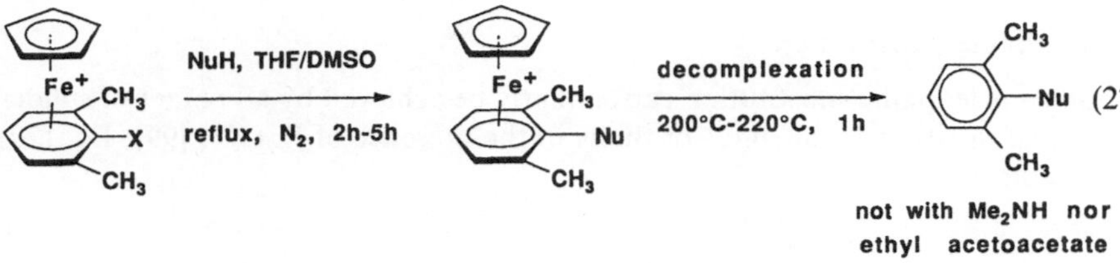

$$\text{not with Me}_2\text{NH nor ethyl acetoacetate} \quad (27)$$

Table 7

	Nu	NH_2	Me_2N	n-BuNH	C_4H_8N*	EtO	PhO	o-$CH_3SC_6H_4$	$CH(COCH_3)COOEt$
Yield %	X = Cl	56	51	68	70	82	69	78	68
	X = NO_2	59	57	69	65	78	65	63	65

* N-pyrrolidyl

The nucleophilic substitution of the nitro group in nitro-arene complexes works almost as well as that of Cl^- and such substitutions were achieved by Chowdhurry et al. with O, S, and N nucleophiles and with stabilized carbanions [97, 98]; Eq. (28) and Table 8.

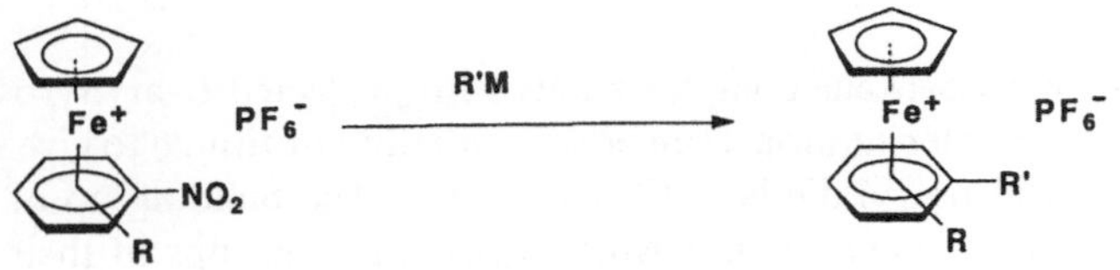

$$(28)$$

Table 8

FeCp(Ar)R, R =	H		o-Me		m-Me		p-Me	
Nucleophile	Nucleofuge							
R' in R'M	Cl	NO$_2$	Cl	NO$_2$	Cl	NO$_2$	Cl	NO$_2$
OCH$_2$CF$_3$	65	75					76	85
CH(COMe)$_2$	76	67						
CH$_2$COCH$_3$	75	65						
CH(CO$_2$Et)$_2$		75						
CH(CO$_2$Et)COCH$_3$		71						
MeO	60		65		62		67	
EtO	64		56		50		60	
p-CH$_3$C$_6$H$_4$S	66		83		84		70	

5.2 Dichloroaromatics

5.2.1 Double Substitution

Double nucleophilic substitution can similarly be achieved by alkoxides of sufides generated in situ from alcohols of thiols in the presence of K$_2$CO$_3$ [99]; Eq. (29) and Table 9.

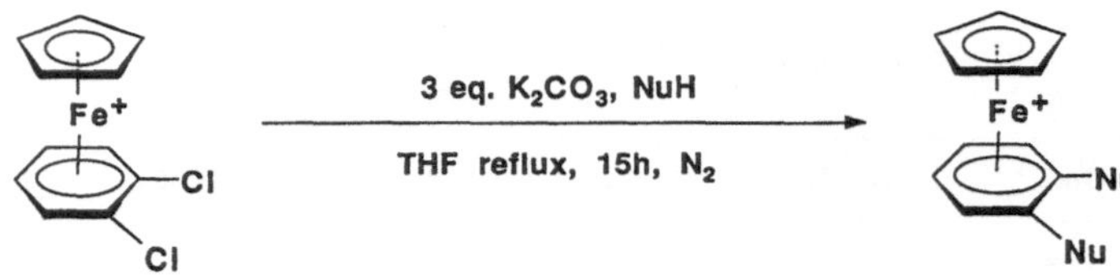

$$\text{(29)}$$

Table 9

Nu	Yield %
PhO	76
p-CH$_3$C$_6$H$_4$S	80
PhCH$_2$O	58
CH$_3$O	80

The *meta*-dichlorobenzene complex reacts with protected *O*-aryltyrosines to give aryl ethers. Both chlorine atoms can be sequentially substituted to give symmetrical or disymmetrical triaryl diethers (Scheme XVI). The building up of such diaryl ethers from phenolic compounds which have amino groups in their side chains

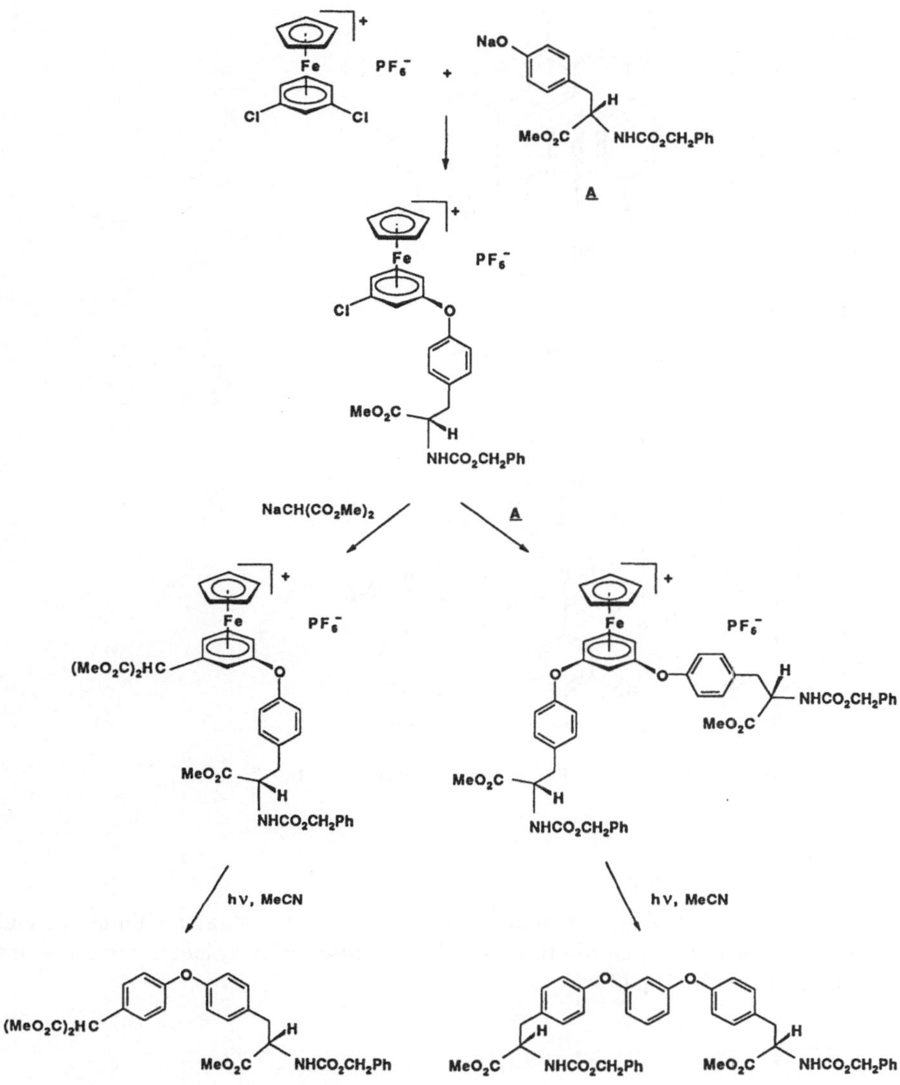

Scheme XVI

is a key step in the synthesis of cyclic antibiotics such as OF4949. I–IV, bouvardin, K-3, and the glycopeptids vancomycins and ristocetine [100]; Fig. 7. The decomplexation was cleanly achieved using visible light irradiation in acetonitrile, a method which has been used for a decade to isolate sophisticated aromatics after activation by the FeCp$^+$ unit [76]. The strategy was extended to racemic *N*-acetylchlorophenylalanine by temporary complexation to RuCp$^+$ which led to the coupling of aromatic amino acids.

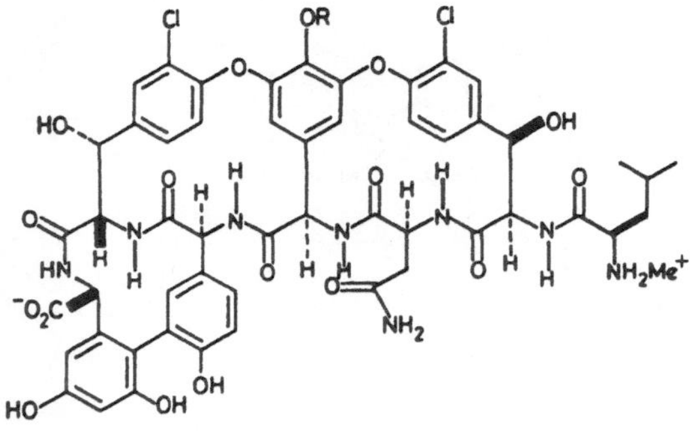
Vancomycin (R = sugar unit)

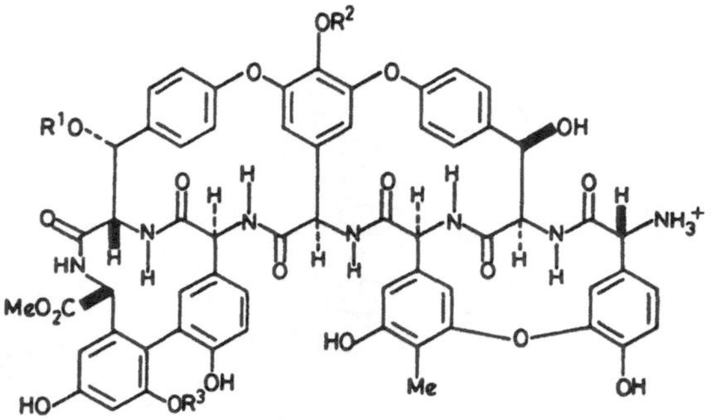
Ristocetine A (R1 - R3 = sugar units)

Fig. 7.

5.2.2 Monosubstitution

In the case described above, monosubstitution can be obtained under careful conditions including high dilution and slow addition of a stoichiometric amount of the nucleophile; Eq. (30):

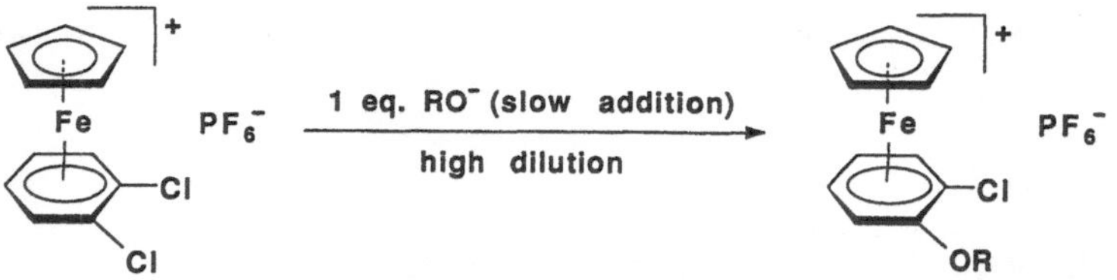
1 eq. RO⁻ (slow addition)
high dilution

(30)

In the reaction of amines such as NH_3, NH_2NH_2, $MeNH_2$, $C_6H_{11}CH_2NH_2$, and $Ph(OMe)NH_2$, only monosubstitution can be obtained even in the presence of a large excess of the amine. This is taken into account by the deprotonation of the acidic monosubstituted complex by free amine leading to an iminocyclohexadienyl complex. The latter cannot be subjected to nucleophilic substitution of the second

Cl substituent. Upon hydrolysis, the monosubstituted iron-arene complex is thus isolated.

When the nucleophile is a stabilized "carbanion" such as the enolate of acetylacetone, 1-benzoylacetophenone, diethylmalonate, or ethyl acetatoacetone, the reaction proceeds similarly. The monosubstituted complex is isolated as long as it contains an acidic hydrogen in the benzylic position. In addition, for the case of diketones $CH_2(COR)_2$ (R = Me, Ph, OEt), a deacetylation is observed in an acidic medium [92, 93]. These features are the same as described above in the case of the substitution of Cl by stabilized carbanions in monochloroaromatics (the second chlorine being an inert arene substituent [99]; Scheme XVII, Eq. (31) and Tables 10 and 11.

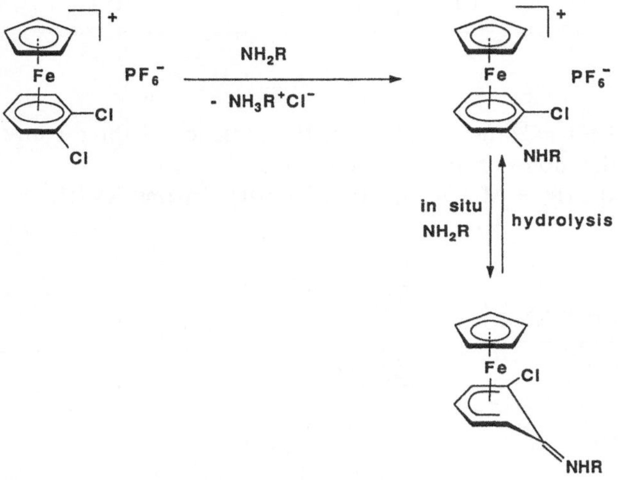

Scheme XVII

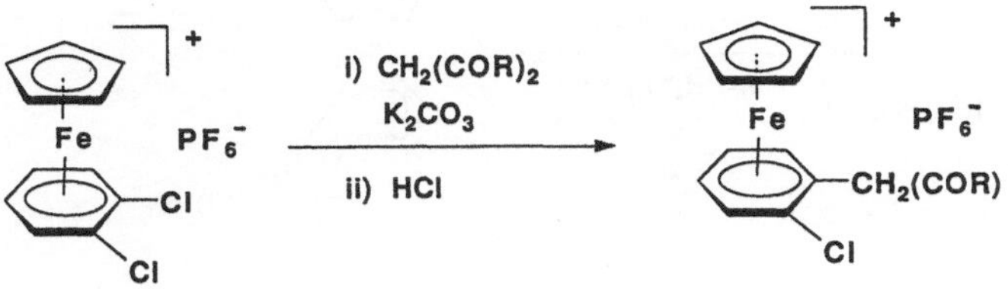

(31)

Table 10

R	Yield %
OEt	65–70
Ph	68–75
CH$_3$	70–80

Table 11. Yield of the monosubstitution of Cl in the complexes
$[o, m,$ and $p\text{-FeCp}(\eta^6\text{-}C_6H_4Cl_2)]^+$

R in reactive R$^-$	Yield %	R in carbanion R$^-$	Yield %
C_6H_5O	68[a]	CH_2COCH_3	81[a]
$p\text{-}CH_3C_6H_4S$	84[a]	CH_2COH_3	75[b]
$C_6H_5CH_2O$	50[a]	CH_2COCH_3	63[c]
CH_3O	75[a]	$CH(COC_6H_5)_2$	70[a]
$o\text{-}CH_3OC_6H_4NH$	60[a]	$CH(COC_6H_5)_2$	68[b]
NH_2	83[a]	$CH(COC_6H_5)_2$	70[c]
NH_2NH	46[a]	$CH(COCH_3)(CO_2C_2H_5)$	80[a]
CH_3NH	85[a]	$CH(COCH)_3(CO_2C_2H_5)$	75[b]
$C_6H_5CH_2NH$	65[a]	$CH(COCH_3)(COOC_2H_5)$	81[c]
		$CH(CO_2C_2H_5)_2$	71[a]

[a] *ortho*; [b] *meta*; [d] *para*

Similarly, the reaction of nitroalkanes at 25 °C in the presence of three equiv.
K_2CO_3 selectively gives the monosubstituted complexes.
$[FeCpC_6H_4(Cl)CHRNO_2]^+$ (R = Me: 74%; Et: 65% [101], Scheme XVIII).

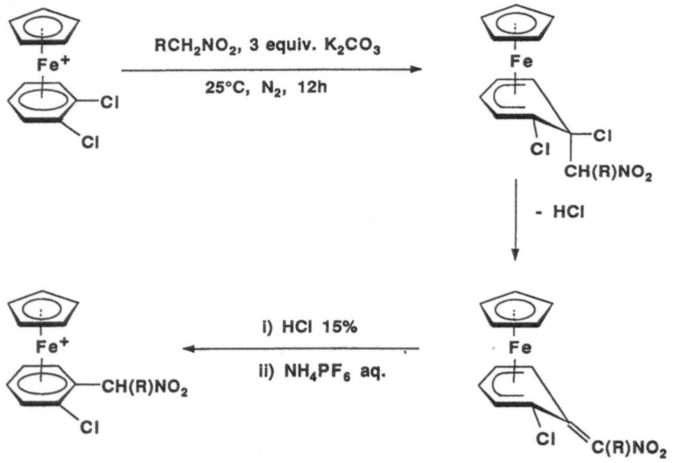

Scheme XVIII

Recently, it was shown that the attack of CN$^-$ on $[FeCp(C_6H_5Cl)]^+PF_6^-$ in DMF
occurs on the *ortho*-position. In the intermediate cyclohexadienyl complex, the
CN group migrates to the *ipso*-carbon, whereas Cl$^-$ is displaced. The monosubstituted benzonitrile complex is subjected to a second *ortho*-CN$^-$ attack but
hydride is not removed spontaneously to give back an arene complex (Scheme
XIX). Removal of the hydride is achieved by oxidation using DDQ (2,3-dichloro-

5,6-dicyano-1,4-benzoquinone) to give the *ortho*-dicyanobenzene complex. The reaction also proceeds similarly with the complexes of PhSO$_2$, *p*-chlorotoluene, and 2,6-dichlorotoluene [102, 103].

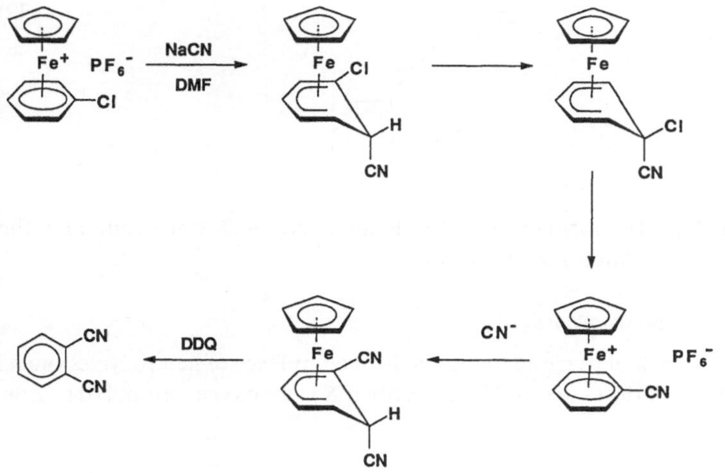

Scheme XIX

5.2.3 Chemistry of Heterocycles

5.2.3.1 Formation of Heterocycles

The double substitution of both chlorine atoms in the complex of *o*-dichlorobenzene can, under certain conditions, lead to the formation of complexes of heterocycles [99, 100, 104]; Scheme XX:

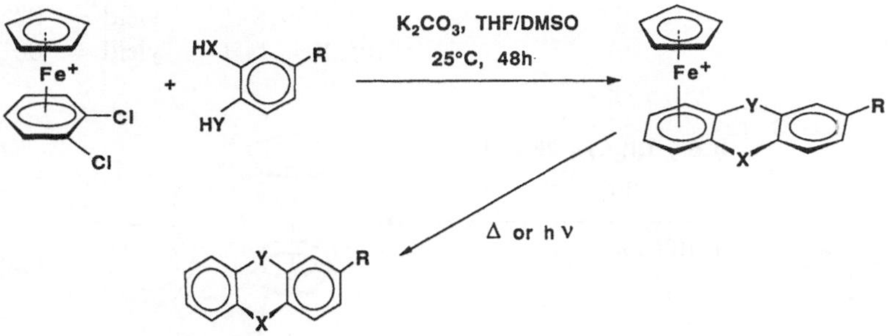

Scheme XX

If X and Y are O or S, the reaction is easy. If X is O or S and Y is NH, the reaction is more difficult. The attack of Y must follow that of X in order to prevent deprotonation (Eq. (32)) leading to an inert monosubstituted complex (as shown above in Scheme XX).

D. Astruc

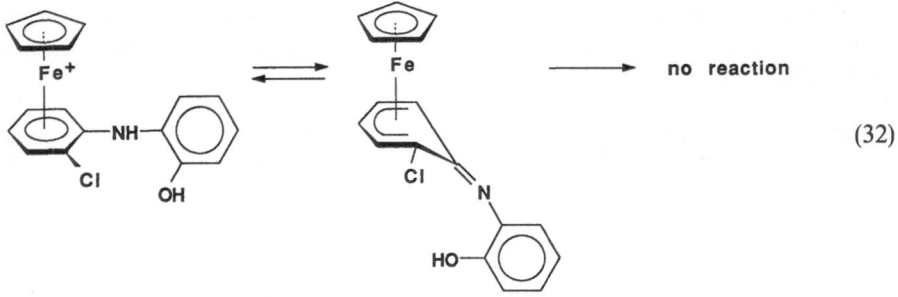

(32)

If X and Y are NH, the formation of the heterocycle does not occur and the reaction stops after the monosubstitution.

5.2.3.2 Opening of Heterocycles

The nucleophilic attack of pyrrolidine on the iron complexes of heterocycles opens the ring by cleavage of the $C-X-C$ bridge only if X is an oxygen atom [104]; Eqs. (33) and (34):

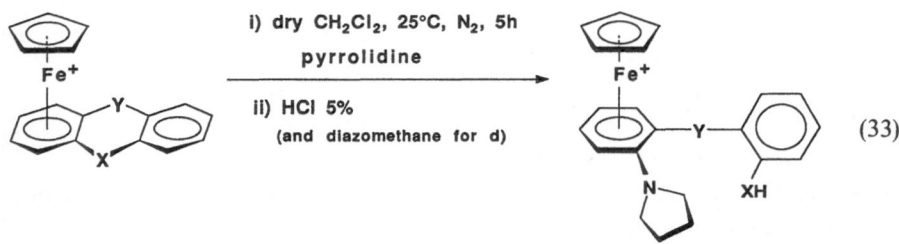

(33)

a) X = O
b) X = S
c) X = NH

a) X = O   yield = 70%
b) X = S   yield = 0%
c) X = NH   yield = 0%

(34)

if X different of O the yield is 0%

a) Y = CH$_2$	R = H	yield = 79%
b) Y = O	R = H	yield = 72%
c) Y = S	R = CH$_3$	yield = 79%
d) Y = NH	R = H	yield = 0%

In order to explain the selective cleavage of the C–O bond, one has to take into account two factors (i) the oxonium is a better leaving group than the sulfonium (ii) for a nucleophilic substitution, weak bases are better leaving groups. The proposed mechanism follows Scheme XXI [99, 104]:

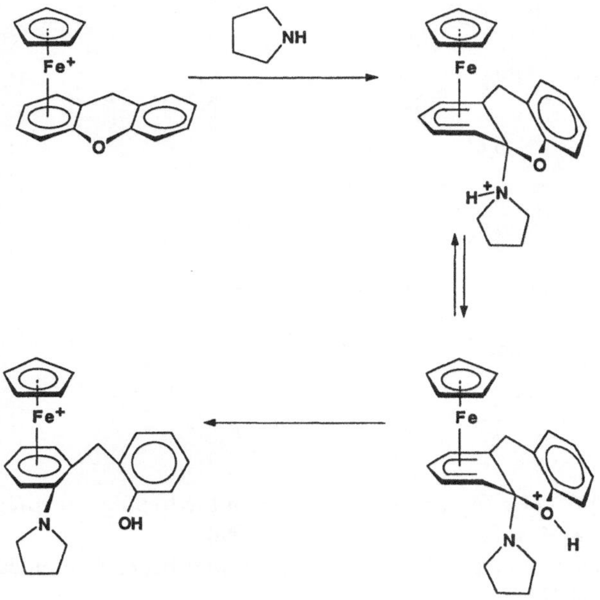

Scheme XXI

6 Redox Reactions of Arene Substituents

These reactions have been reviewed [11, 12] and will be shortly summarized here. Many of them were disclosed in the pioneering work of Nesmeyanov's group [105].

6.1 Oxidation

Since the $FeCp(C_6H_6)^+$ unit is robust towards oxidation even in concentrated sulfuric acid, oxidation of alkyl substituents upon boiling in aqueous $KMnO_4$ solution can be achieved and leads to carboxylic substituents. The mesitylene complex can be oxidized to the mono-, di-, or tri-carboxylic acid depending upon the reaction conditions. In the latter case, the decomplexed trimesic acid is obtained [106, 107]; Scheme XXII:

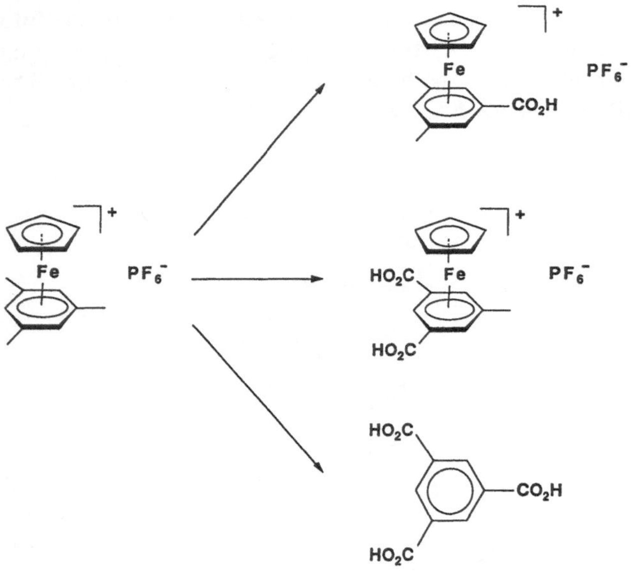

Scheme XXII

As expected, the acid $[FeCp(C_6H_5CO_2H)]^+$ is stronger than the free ligand (pK_a values in 50% aq. ethanol: 3.04 ± 0.4 resp. 5.73 ± 0.04 [108]).

The acids can be esterified or converted to acid chlorides and hence to amides and nitriles [107]; Scheme XXIII:

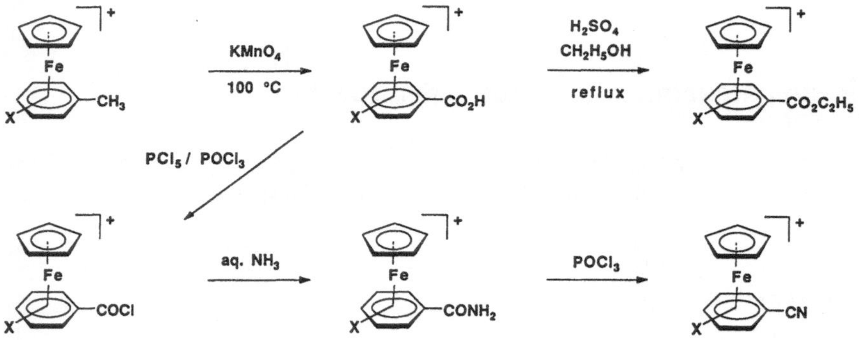

Scheme XXIII

Some alkyl substituents with a secondary benzylic carbon can be oxidized to ketones using $KMnO_4$ at 50 °C in water; Eqqs. (35) and (36). In the case of the fluorene complex only, this oxidation could be alternatively achieved using aqueous

$H_2O_2/NaOH$ at 25 °C [109]. The analogous oxidation of $[Cr(CO_3)(fluorene)]$ uses KO_2 or a base and O_2 [110].

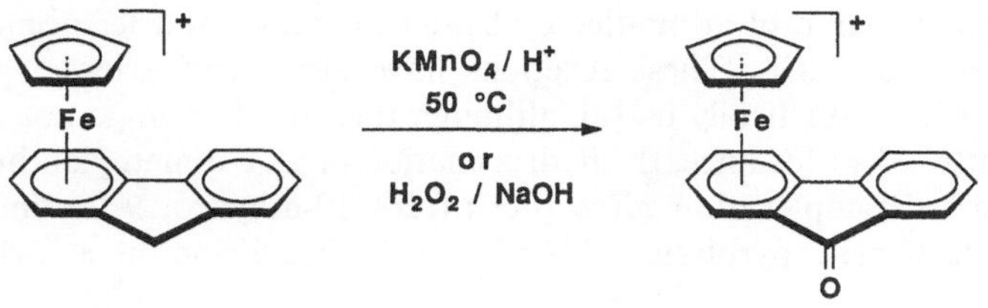

(35)

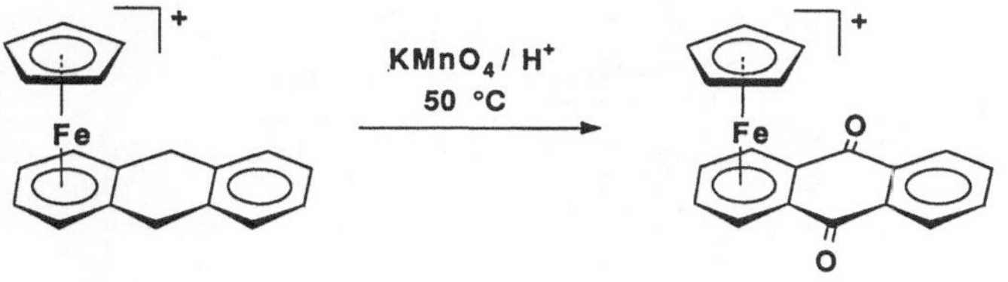

(36)

The thioxanthene ligand is oxidized to thioxanthone and sulfone ligands, the ratio of which depends on the amount of $KMnO_4$; Eq. (37):

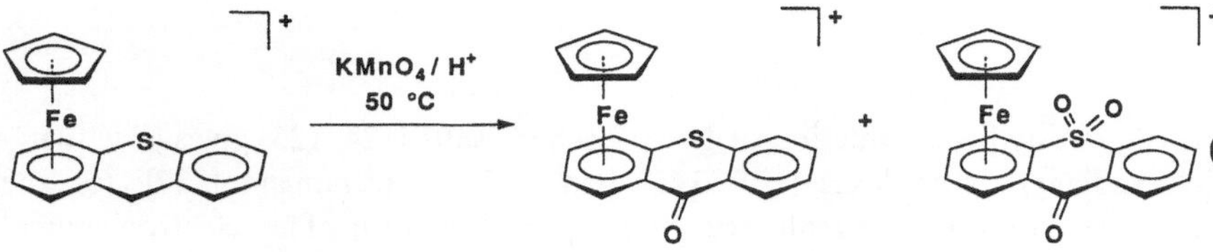

(37)

However, this oxidation to carbonyl failed with the complexes of tetralin, o-chlorotoluene, 9,10-dihydrophenanthrene, and acenaphthalene [109]. The aniline complex can be oxidized to the nitrobenzene complex using H_2O_2 in CF_3CO_2H [86]; Eq. (38). This reaction parallels the analogous oxidation of aminocobalticinium [86, 111].

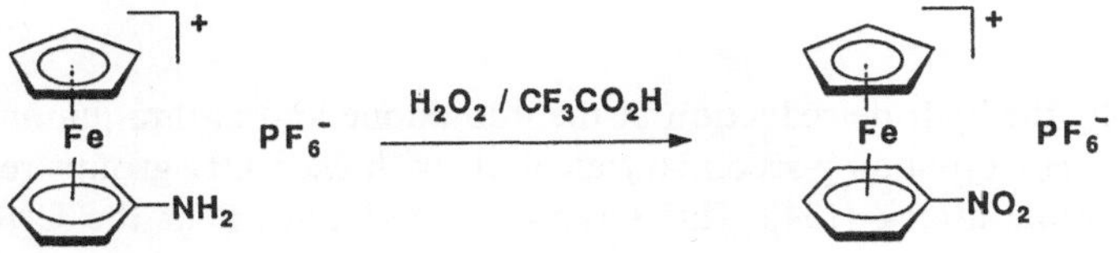

(38)

6.2 Reduction

Hydrogenation of condensed polyaromatics [112–122], dehalogenation (F < Cl < Br ~ I), and dealkylation (Et groups [80, 123]) occur during the course of the complexation of aromatics by ligand exchange of a ferrocene ring under the influence of $AlCl_3$. These reactions have also been reviewed [12] and can sometimes be synthetically useful, although their mechanism is not clear (except for the latter). For instance, the hydrogenation of 9,10-dimethylanthracene in the course of its complexation gives [FeCp(*cis*-9,10-dihydro, 9,10-dimethylanthracene)] in 25% yield (pyrolysis yields 80% of the free ligand there of [115]; Scheme XXIV:

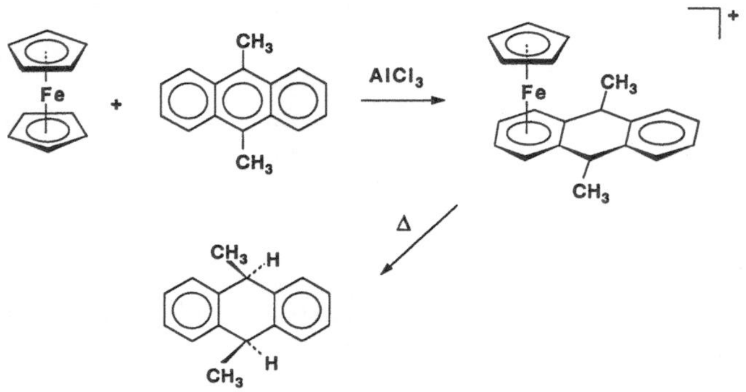

Scheme XXIV

However, the hydride reduction of FeCp(arene)$^+$ salts [124, 125] gives [FeCp(η^5-cyclohexadienyl)] complexes [125, 126] (via an ET mechanism [127]; for the directing effect of substituents see Refs. [126, 128–130]. The electrochemical reduction of the carboxylic substituents at an Hg cathode in water leads to the primary alcohol [131–133]; Eq. (39):

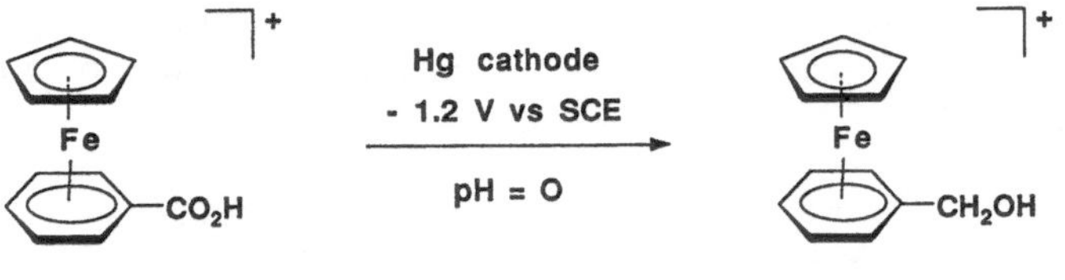

$$(39)$$

Similarly, the hydride reduction of the fluorenone and anthraquinone complexes gives the corresponding secondary alcohols with *endo*-OH groups resulting from stereospecific attack [134]. This strategy is also known in the $Cr(CO)_3$(arene) chemistry and led to optically active aromatic compounds [135]. Later, the electrochemical reduction of the side chain was also extended to the oximes [136]; Eq. (40):

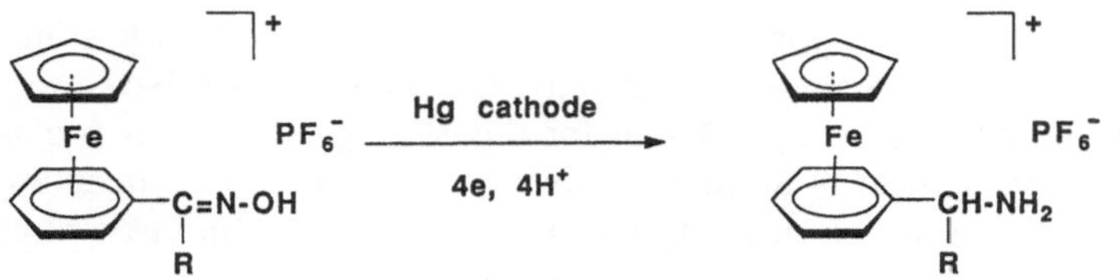

7 Comparison with other Metals. Conclusion and Prospects

Temporary complexation of aromatics to cationic organoiron moieties is an easy way to invert their reactivity. Formation of C–C bonds was an objective of this aromatic chemistry. It could be performed in the bis(arene)iron series either by nucleophilic attack of carbanions on $18e$ complexes or by reaction of halides on $18e$ deprotonated complexes or an the $20e$ complex by an ET path. In the FeCp(arene)$^+$ series, one-pot multiple C–C bond formation can be achieved by mixing an excess of a base and a halide with the polymethylbenzene complex. These reactions lead to tentacled iron sandwiches which can be suitably modified to provide discoid molecules and surfactants with a redox active Fe^I/Fe^{II} center. These new molecules are clearly potential candidates for taking part in processes which require specific physico-chemical properties in solution and in the solid state. Not only deprotonation/alkylation reactions, but also nucleophilic substitutions of chloride by RO$^-$, RS$^-$, NR$_2^-$, and stabilized carbanions including CN$^-$ are powerful ways to form C–C and other C–X bonds (X = O, S, N). The extent and limits of substitution of two chlorides are now understood and can lead to heterocycles under suitable conditions. These trends can now lead to new applications as recently shown in the formation of triaryldiethers. The main types of activation, well known with the Cr(CO)$_3$ group (nucleophilic addition and substitution, deprotonation), now have also been achieved with FeCp$^+$ which is a stronger activator (equivalent to two nitro groups). Complementary metals are Cu, Ni, and Pd which can catalytically activate aryl-halogen bonds with CO insertion [137–142]. The direct complexation by iron of aromatics bearing heteroatom substituents is not possible because of their complexation by AlCl$_3$ inhibiting the synthesis of the iron-arene complex. However, the introduction of such substituents is easy after complexation by the mild nucleophilic substitution procedure. Alternatively, photolytic transmetallation of aromatics bearing heteroatoms is possible and directly leads to FeCp(arene)$^+$ complexes which are not accessible by the Fischer-Haffner-type route [143–146]. The easy complexation by RuCp$^+$ [147–149] or Ru(C$_5$Me$_5$)$^+$ [150–153] now provides an expensive means for the complexation of such aromatics and the activation is somewhat lower. The activation by Mn(CO)$_3^+$ is extremely powerful as already noted in pioneering studies [154–156]. Recent studies have confirmed the potential for synthetic applications of this unit despite its high cost. Reactions of such aromatic complexes with glycine enolate equivalents gave neutral dienyl Mn complexes which were converted to methyl esters of arylglycine. The decomplexation using NBS led to the aromatic derivatives. Aryl oxides also react with $Mn(CO)_3(ArCl)^+$ to give nucleophilic substitution of the chlorine atom. Diarylethers bearing amino ester groups on both rings can be obtained in this way [157]. However, the authors later switched to the $CpFe^+$ unit for this strategy [100]. It is probable that the recent development of aromatic activation by cationic iron moieties will lead to many applications given the lack of toxicity, low cost, and ease of complexation and decomplexation of the iron sandwich units.

Acknowledgements. For their unvaluable help and expertise in the preparation of this review, I am deeply indebted to Dr. Nicole Ardoin and Jocelyne Moncada. Thanks are also due to the students, post-doctoral fellows, and colleagues cited in references who have contributed to the inspiration and achievement of a large proportion of the organoiron chemistry presented therein. In particular, the collaborations with Drs. Jean-René Hamon, Alex Madonik, Dominique Mandon, Françoise Moulines, Loïc Toupet, and Enrique Román were essential. Professor Roald Hoffmann and Dr. Jean-Yves Saillard have considerably helped us in understanding the orbital-controlled regioselectivity of the nucleophilic attacks. Their continuous interest and discussion are gratefully acknowledged. Let us also encourage Laurent Djakovitch, Dr. Jean-Luc Fillaut, and Bruno Gloaguen in their hard work and enthusiasm to achieve the multifunctional tentacled sandwich chemistry in progress and briefly introduced here. This review was written at the Technical University of Munich-Garching in 1989–1990. It is my pleasure to thank Professor Dr. Wolfgang A. Herrmann and his healthy research group and colleagues of the Inorganic Department for a kind hospitality and stimulating interactions. Finally, the Alexander von Humboldt Foundation is gratefully acknowledged for a generous Award. The Centre National de la Recherche Scientifique, the Universities of Bordeaux and Rennes, and the Région Aquitaine are also thanked for their support during the last decade.

8 Abbrevations

A	19e organoiron radical which does not bear benzylic hydrogens
Ar	aryl
BH	19e organoiron radical bearing a benzyclic hydrogen (the latter becomes acidic in the oxidized form BH^+)
9-BBN	9-borabicyclo[3.3.1]nonane
Cp	η^5-cyclopentadienyl
Cp*	η^5-pentamethylcyclopentadienyl
DDQ	2,3-dichloro-5,6-dicyano-1,4-benzoquinone
DMF	dimethylformamide (solvent)

DMSO	dimethylsulfoxide (solvent)
e^-	electron
$12e$	12-electron
$17e$	17-electron
$18e$	18-electron
$19e$	19-electron
$20e$	20-electron
E^0	thermodynamic redox potential
ESR	electron spin resonance
ET	electron-transfer
Et	ethyl
FE	η^5-C_5H_5Fe (sandwich iron complex)
hexaolefin	molecules in the top part of Scheme XIV with six double bonds
hexol	central sandwich molecule of Scheme XII with six OH groups
Me	methyl
NMR	nuclear magnetic resonance
Ph	phenyl
ppm	part per million
resp	respectively
RM	organometallic reagent
S	reducible organic substrate
t-Bu or *t*-butyl	*tert*-butyl
THF	tetrahydrofurane (solvent)
TMS	tetramethylsilane (internal NMR reference for δ)
UV	ultra-violet light
V	Volts
vs	versus
δ	NMR chemical shift with TMS as the reference

9 References

1. a) Keally TJ, Pauson PJ (1951) Nature (London) 168: 1039
   b) Miller SA, Tebboth JA, Tremaine JF (1952) J Chem Soc 632
2. a) Wilkinson G, Rosenblum M, Whitting MC, Woodward RB (1952) J Am Chem Soc 74: 2125
   b) Wilkinson G (1975) J Organomet Chem 100: 273
3. Fischer EO, Pfab W (1952) Z Naturforsch (B 7): 377
4. Wilkinson G, Stone FGA, Abel WE (eds) (1982) Comprehensive organometallic chemistry (vols 4 and 8) Pergamon Press, Oxford
5. King RB (1978) In: The organic chemistry of iron, Koerner von Gusthorf EA, Grevels FW, Fischler I (eds), vol 1, Academic Press, N.Y., p 525
6. Rosenblum M (1965) Chemistry of the iron group metallocenes (part I) Wiley, New York
7. a) Becker EI, Tsutsui M (1972) Organometallic reactions vol 4, Wiley, New York, p 163
   b) Black DStc, Jackson WR, Swan JM (1979) In: Comprehensive organic chemistry, Barton D, Ollis WO (eds), vol 3, Pergamon Press, New York, Chap 156, p 1127
   c) Knox GR, Watts WE (1976) In: International review of science, inorganic chemistry, Series 2. vol 6, Transition Metals. Mays MJ (ed) Butterworth, London, p 219

    d) Jolly WL (1968) Inorg synth 11: 120

    e) Rinehart KL (1969) Organic reactions vol 17, Wiley, New York

    f) Rinehart KL Gmelin Handbuch der Anorganischen Chemie. vol 49, part A

8. a) Green MLH (1968) Organometallic compounds vol II: The transition elements, Methuen, London

    b) Collman JP, Hegedus LS (1980) Principles and applications of organotransition metal chemistry, University Science Books, Mill Valley, CA

    c) Cotton FA, Wilkinson G (1980) Advanced inorganic chemistry, 4th edn. Wiley, New York

9. a) Semmelhack MF (1977) Annals N.Y. Acad Sci 295: 36

    b) Jaouen G (1977) Annals N.Y. Acad Sci 295: 59

10. Sutherland RG (1977) J Organomet Chem Library 3: 311; see also ref. 96

11. Astruc D (1983) Tetrahedron Report n° 157, Tetrahedron 39: 4027

12. Astruc D (1987) In: Hartley FR, Patai S (eds) The chemistry of the metal-carbon bond, Wiley, New York, vol IV, p 625

13. Balas L, Jhurry D, Latxague L, Grelier S, Morel Y, Hamdani M, Ardoin N, Astruc D (1990) Bull Soc Chim Fr, 127: 401

14. Kane-Maguire LAP, Honig ED, Sweigart DA (1984) Chem Rev 84: 525

15. Eberson L (1987) Electron-transfer reactions in organic chemistry, Springer-Verlag, Berlin, Heidelberg, New York, p 172

16. Chanon M (1987) Acc Chem Res 20: 214

17. Savéant JM (1980) Acc Chem Res 13: 323

18. Astruc D (1988) Chem Rev 88: 1189

19. a) Nesmeyanov AN (1968) Int Union Pure Applied Chemistry, Butterworth, London, p 221

    b) Nesmeyanov AN (1972) Adv Organomet Chem 10: 1

20. a) Fischer EO, Böttcher R (1956) Chem Ber 89: 2397

    b) Helling JF, Braitsch DM, Meyer TJ (1971) J Chem Soc Chem Commun 920

21. Astruc D (1986) Acc Chem Res 19: 377

22. Astruc D (1988) Angew Chem Int Ed 27: 643, 126 ref, (1988) Angew Chem 100: 662

23. Mandon D, Astruc D (1989) J Organomet Chem 369: 383

24. Hamon JR, Astruc D (1988) Organometallics 8: 2372

25. Fischer EO, Röhrscheid F (1962) Z Naturforsch B 17: 483

26. Rajasekharan MV, Giezynski S, Ammeter JH, Ostwald N, Michaud P, Hamon J-R, Astruc D (1982) J Am Chem Soc 104: 2400

27. Brintzinger H, Palmer G, Sands RH (1966) J Am Chem Soc 88: 623

28. Michaud P, Mariot JP, Varret F, Astruc D (1982) J Chem Soc Chem Commun 1383

29. Weber SE, Brintzinger HH (1977) J Organomet Chem 127: 45

30. Cameron TS, Clerk MD, Linden A, Sturge K, Zawarotko MJ (1988) Organometallics 7: 2571

31. a) Helling JF, Braitsch DM (1970) J Am Chem Soc 92: 7207

    b) Helling JF, Rice SL, Braitsch DM, Mayer T (1970) J Am Chem Soc 92: 7209

32. Helling JF, Cash GG (1974) J Organomet Chem 73: C10

33. Davies SG, Green MLH, Mingos DMP (1978) Tetrahedron Report n° 57, Tetrahedron 34: 3047

34. Davies SG (1982) In: Organotransition metal chemistry: application to organic synthesis, Pergamon Press, Oxford

35. Madonik AM, Mandon D, Michaud P, Lapinte C, Astruc D (1984) J Am Chem Soc 106: 3381

36. Astruc D, Michaud P, Madonik AM, Saillard JY, Hoffmann R (1985) Nouv J Chim 9: 41

37. Nesmeyanov AN, Volkenau NA, Bolesova VN, Polkonikova LS (1977) Dokl Akad, Nauk SSSR 236: 1130

38. Pearson AJ (1980) Acc Chem Res 13: 463

39. a) Mandon D, Toupet L, Astruc D (1986) J Am Chem Soc 108: 1320

    b) Mandon D, Astruc D (1989) Organometallics 8: 1372

40. Lamana W, Brookhart M (1981) J Am Chem Soc 103: 989

41. Brookhart M, Lamana W, Pinkas AP (1983) Organometallics 2: 638
42. a) Brookhart M Lukacs A (1984) J Am Chem Soc 106: 4161
    b) Brookhart M, Lamana W, Humphrey MB (1982) J Am Chem Soc 104: 2117
43. Chung Y, Khoi HS, Sweigart DA, Connelly NG (1982) J Am Chem Soc 104: 4245
44. Chung Y, Sweigart DA, Connelly NG, Sheridan JB (1985) J Am Chem Soc 107: 2388
45. Pike RD, Ryan WJ, Carpenter GB, Sweigart DA (1989) J Am Chem Soc 111: 8535
46. a) Mandon D, Astruc D (1986) J Organomet Chem 307: C27
    b) Mandon D, Astruc D (1990) Organometallics 9: 341
47. a) Madonik AM, Astruc D (1984) J Am Chem Soc 106: 2437
    b) Astruc D, Mandon D, Madonik AM, Michaud P, Varret F, Ardoin N (1990) Organometallics 9: 2155
48. Nesmeyanov AN, Vol'kenaau NA, Petrakova VA (1977) J Organomet Chem 136: 363
49. a) Sawyer DT, Gibian MJ (1979) Tetrahedron 35: 1471
    b) Sawyer DT (1979) Acc Chem Res 3: 105
50. Sawyer DT, Gibian MJ, Morrison MM, Sev ET (1978) J Am Chem Soc 100: 627
51. Ruff E (1977) Chem Soc Review 6: 195
52. Michelson AM, McCord JM, Fridovich I (1977) In: Superoxide and superoxide dismutase. Academic Press, New York
53. Fridovich I (1976) In: Pryor WA (ed) Free radicals in biology. Academic Press, New York, p 239
54. Astruc D, Román E, Hamon JR, Batail P (1979) J Am Chem Soc 101: 2240
55. Hamon JR, Astruc D, Román E, Batail P, Mayerle JJ (1981) J Am Chem Soc 103: 2431
56. Michaud P, Astruc D (1982) J Chem Soc, Chem Commun 416; see also ref. 73.
57. Astruc D, Hamon JR, Román E, Michaud P (1981) J Am Chem Soc 103: 7502
58. Hamon JR, Astruc D (1988) Organometallics 7: 1036
59. Sawyer DT, Valentine JS (1981) Acc Chem Res 14: 393
60. Lacoste M, Ardoin N, Astruc D, unpublished work
61. Trahanovsky WS, Card RJ (1972) J Am Chem Soc 94: 2897
62. Jaouen G, Meyer A, Simmoneaux G (1985) J Chem Soc, Chem Commun 813
63. Pauson PL, Segal JA (1975) J Chem Soc, Dalton Trans 1677
64. a) Johnson JW, Treichel PM (1976) J Chem Soc, Chem Commun 688; (1977) J Am Chem Soc 99: 1427
65. Lee CC, Steele BR, Demchuck KJ, Sutherland RG (1979) Can J Chem 57: 946; (1979) J Organomet Chem 181: 411
66. a) Helling JF, Hendrickson WA (1977) J Organomet Chem 141: 99
67. Johnson JW, Treichel PM (1972) J Chem Soc, Chem Commun 688
68. Johnson JW, Treichel PM (1977) J Am Chem Soc 99: 1427
69. Catheline D, Astruc D (1984) Nouv J Chim 8: 381
70. Ustynyuk NA, Denisovitch LI, Peterleitner MB, Novikova LN, Pomazanova NA, Kravtsov DN (1988) Metalloorg Khim 1: 216
71. Lee CC, Gill RG, Sutherland RG (1981) J Organomet Chem 206: 89
72. Helling JF, Hendrickson WA (1979) J Organomet Chem 168: 87
73. Moinet C, Raoult E (1982) J Organomet Chem 229: C13; 231: 245
74. Hamon JR, Guénot P, Sinbandhit S, Hamon P, Astruc D, J Organomet Chem 1991, in press
75. Astruc D, Hamon JR, Althoff G, Román E, Batail P, Michaud P, Mariot JP, Varret F, Cozak D (1979) J Am Chem Soc 101: 5445
76. a) Hamon JR, Saillard JY, Le Beuze A. McGlinchey M, Astruc D (1982) J Am Chem Soc 104: 7549
    b) Fillaut JL, Astruc D (work in progress)
77. Moulines F, Astruc D (1988) Angew Chem 100: 1394; (1988) Angew Chem Int Ed Engl 27: 1347
78. Moulines F, Toupet L, Astruc D, to be submitted for publication
79. Dubois RH, Zaworotko MJ, White PS (1989) J Organomet Chem 362: 155
80. Hamon JR, Saillard JY, Toupet L, Astruc D (1989) J Chem Soc, Chem Commun 1662

81. a) Moulines F, Astruc D (1989) J Chem Soc, Chem Commun 614
    b) Moulines F, Djakovitch L, Astruc D, work in progress
82. Djakowitch L, Fillaut JL, Astruc D, unpublished work
83. Nesmeyanov AN, Vol'kenau NA, Bolesova IN (1967) Dokl Akad Nauk SSSR 175: 606
84. Nesmeyanov AN, Vol'kenau NA, Isaeva LS, Bolesova IN (1968) Dokl Akad Nauk SSSR 183: 354
85. Nesmeyanov AN, Vol'kenau NA, Isaeva LS (1967) Dokl Akad Nauk SSSR 176: 106
86. Lee CC, Gill US, Iqbal M, Azogu CI, Sutherland RG (1982) J Organomet Chem 231: 151
87. Gill US (1982) PhD Thesis, Univ Saskatchevan (Canada)
88. Fillaut J-L, Astruc D (unpublished work)
89. Lee CC, Azogu CI, Chang PC, Sutherland RG (1981) J Organomet Chem 220: 181
90. Sutherland RG, Chang PC, Lee CC (1982) J Organomet Chem 234: 197
91. Vichi EJS, Miller J, Moran PS, Gornes PCB XIII Intern Conf Organomet Chem, Turin, Sept 1988, Abstr 310
92. Moriarty RM, Gill US (1986) Organometallics 5: 253
93. Gill US, Moriarty RM (1986) Synth React Inorg Met Org Chem 16: 485
94. Khand IU, Pauson WE, Watts WE (1969) J Chem Soc, (C) 2024
95. Nesmeyanov AN, Vol'kenau NA, Isaeva LS (1967) Dokl Akad Nauk SSSR 176: 106
96. Abd-El Aziz AS, Lee CC, Piorko A, Sutherland RG (1988) J Organomet Chem 348: 95
97. Lee CC, Abd-El Aziz AS, Chowdhurry RL, Piorko A, Sutherland RG (1986) Synth React Inorg Met Org Chem 16: 541
98. Chowdhurry RL, Lee CC, Piorko A, Sutherland RG (1985) Synth React Inorg Met Org Chem 15: 1237
99. Lee CC, Abd-El Aziz AS, Chowdhurry RL, Gill US, Piorko A, Sutherland RG (1986) J Organomet Chem 315: 79
100. Pearson AJ, Park JG, Yang SH, Chuang YH (1989) J Chem Soc, Chem Commun 1363
101. Gill US, Moriarty RM (1986) Synth React Inorg Met Org Chem 16: 1103
102. Sutherland RG, Chowdhury RL, Piorko A, Lee CC (1987) J Organomet Chem 319: 379
103. Sutherland RG, Zhang CH, Piorko A, Lee CC (1989) Can J Chem 67: 137
104. a) Lee CC, Piorko A, Steele BR, Gill US, Sutherland RG (1983) J Organomet Chem 256: 303
     b) Sutherland RG, Piorko A, Gill US, Lee CC (1982) J Heterocyclic Chem 19: 801
105. a) Nesmeyanov AN (1968) Int Union Pure Applied Chemistry, Butterworth, London, p 221
     b) Nesmeyanov AN (1972) Adv Organomet Chem 10: 1
106. Nesmeyanov AN, Vol'kenau NA, Sirotkina EI, Deryabin VV (1967) Dolk Chem 177: 1110
107. Nesmeyanov AN, Vol'kenau NA, Sirotkina EI (1967) Izv Akad Nauk SSSR, Ser Khim 1170
108. Sirotkina EI, Nesmeyanov AN, Vol'kenau NA (1968) Izv Akad Nauk SSSR, Ser Khim 1605
109. Sutherland RG, Gill US, 180th Nat Meeting American Chemical Society, Las Vegas, Aug 1980
110. Top S, Jaouen G, McGlinchey M (1980) J Chem Soc, Chem Commun 643
111. Sheats JE, Rausch MD (1970) J Org Chem 35: 3245
112. Lee CC, Sutherland RG, Thompson BJ (1971) J Chem Soc, Chem Commun 1071
113. Sutherland RG, Chen SC, Pannekoek WJ, Lee CC (1975) J Organomet Chem 101: 221
114. Sutherland RG, Chen SC, Pannekoek WJ, Lee CC (1976) J Organomet Chem 117: 61
115. Sutherland RG, Pannekoek WJ, Lee CC (1977) Annals N.Y. Acad Sci 295: 192
116. Sutherland RG, Pannekoek WJ, Lee CC (1978) Can J Chem 56: 1782
117. Lee CC, Demchuk KJ, Pannekoek WJ, Sutherland RG (1978) J Organomet Chem 162: 253
118. Lee CC, Demchuk KJ, Sutherland RG (1979) Can J Chem 57: 933
119. Lee CC, Steele KJ, Sutherland RG (1980) J Organomet Chem 186: 265
120. Lacoste M, Rabaa H, Astruc D, Le Beuze A, Scillard JY, Précigona G, Courseille C, Ardvin N, Bouyer W (1989) Organometallics 8: 2233

121. Guerchais V, Astruc D (1983) J Chem Soc, Chem Commun 1115
122. Guerchais V, Astruc D (1986) J Organomet Chem 312: 97
123. Hamon JR, Astruc D (1989) Organometallics 8: 2243
124. Green MLH, Pratt L, Wilkinson G (1960) J Chem Soc 989
125. Jones D, Pratt L, Wilkinson G (1962) J Chem Soc 4458
126. Khand IU, Pauson PL, Watts WE (1968) J Chem Soc C 2257
127. Michaud P, Astruc D, Ammeter JH (1982) J Am Chem Soc 104: 3755
128. Khand IU, Pauson PL, Watts WE (1969) J Chem Soc 116
129. Khan MM, Watts WE (1976) J Organomet Chem 108: C11
130. McGreer JF, Watts WE (1976) J Organomet Chem 110: 103
131. Román E, Dabard R, Moinet C, Astruc D (1979) Tetrahedron Letters 16: 1433
132. Román E, Astruc D, Darchen A (1976) J Chem Soc, Chem Commun 183
133. Román E, Astruc D, Darchen A (1981) J Organomet Chem 219: 221
134. Lee CC, Demchuk KJ, Gill US, Sutherland RG (1983) J Organomet Chem 247: 71
135. Jaouen G, Meyer A (1975) J Am Chem Soc 97: 4667
136. Moinet C, Raoult E (1982) J Organomet Chem 229: C13
137. Sekiya A, Ishikawa N (1975) Chem Lett 277
138. Ben David Y, Portnoy M, Milstein D (1989) J Am Chem Soc 111: 8742
139. Tagasaki K, Okamoto T, Sakakibara Y, Ohno A, Oka S, Hayama N (1975) Bull Chem Soc Jpn 48: 3298
140. Tagaski K, Okamoto T, Sakakibara Y, Ohno A, Hayama N (1976) Bull Chem Soc Jpn 49: 3177
141. Cassar L, Ferra S, Foa S (1974) Homogeneous Catalysis-11, Advances in chemistry series, American Chemical Society, Washington 132: 252
142. Mutin R, Lucas C, Thiovolle-Cazat J, Dufaud V, Dany F, Basset JM (1988) J Chem Soc, Chem Commun 896
143. Gill TP, Mann KR (1980) Inorg 19: 3008; (1981) J Organomet Chem 216: 65
144. Catheline D, Astruc D (1983) J Organomet Chem 248: C9
145. Laganis ED, Finke RG, Boekelheide V (1981) Proc Natl Acad Sci USA 78: 2657
146. Román E, Astruc D (1986) Bol Soc Chil Quim 31: 129
147. Gill TP, Mann KR (1982) Organometallics 1: 485
148. Moriarty RM, Yi-Yin Ku, Gill US (1987) J Chem Soc, Chem Commun 1493; (1990) Ibid 1764 (steroids).
149. Segal JA (1985) J Chem Soc, Chem Commun 1338
150. Moriarty RM, Yi-Yin Ku, Gill US (1987) J Chem Soc, Chem Commun 1837.
151. Fagan PJ, Ward MD, Calabrese JC (1989) J Am Chem Soc 111: 1698
152. a) Koëlle V, Kossakowski J (1988) J Chem Soc, Chem Commun 549; (1989) J Organomet Chem 362: 383
    b) Koëlle V, Wang MH (1990) Organometallics 9: 195
153. a) Chaudret B, He X, Huang Y (1989) J Chem Soc, Chem Commun 1844
    b) Chaudret B, Jalon F (1988) J Chem Soc, Chem Commun 711
154. Pauson PL, Segal JA (1975) J Chem Soc Dalton Trans 1677
155. Watts WE (1982) In: Comprehensive organometallic chemistry, Pergamon Press, New York, vol 8, p 1013
156. Brown DA, Raju JR (1966) J Chem Soc A 40
157. Pearson AJ, Bruhn PR, Gonzales F, Lee SH (1989) J Chem Soc, Chem Commun 659; (1986) J Org Chem 51: 2137

Transition Metal Complexes of Sterically Demanding Cyclopentadienyl Ligands

Jun Okuda

Anorganisch-chemisches Institut der Technischen Universität München, Lichtenbergstraße 4, D-8046 Garching, FRG

Table of Contents

Topics in Current Chemistry, Vol. 160
© Springer-Verlag, Berlin Heidelberg 1991

List of Abbreviations

Cp:	η^5-cyclopentadienyl
Cp′:	η^5-methylcyclopentadienyl
Cp*:	η^5-pentamethylcyclopentadienyl
BuCp:	η^5-*tert*-butylcyclopentadienyl
Bu_2Cp:	η^5-1,3-di-*tert*-butylcyclopentadienyl
Bu_3Cp:	η^5-1,2,4-tri-*tert*-butylcyclopentadienyl
Pr_4Cp:	η^5-tetra-*iso*-propylcyclopentadienyl
Pr_5Cp:	η^5-penta-*iso*-propylcyclopentadienyl
SiCp:	η^5-trimethylsilylcyclopentadienyl
Si_2Cp:	η^5-1,3-bis(trimethylsilyl)cyclopentadienyl
cod:	1,5-cyclooctadiene
dmpe:	1,2-bis(dimethylphosphino)ethane
Si_3Cp:	η^5-1,2,4-tris(trimethylsilyl)cyclopentadienyl
$BuSi_2Cp$:	η^5-4-*tert*-butyl-1,2-bis(trimethylsilyl)cyclopentadienyl

The present review summarizes recent developments in the design of selected cyclopentadienyl ligands that contain at least two sterically demanding substituents at the ring periphery and the use of these ligands in organotransition metal chemistry. General features concerning ligand synthesis, complexation reactions, structure, and reactivity of the complexes as well as their application are reviewed. Special emphasis is put on the description of unusual interligand interactions, conformational analysis, and the use as stoichiometric reagents and catalyst precursors in transition metal-mediated transformation of organic molecules.

1 Introduction

Ever since the serendipitous synthesis of bis(η^5-cyclopentadienyl)iron or ferrocene $FeCp_2$ exactly forty years ago [1] and the subsequent characterization of the structure and metal-ligand bonding of this prototypal sandwich complex [2], the parent cyclopentadienyl ligand Cp has become one of the most important and widely used ligands not only in organotransition metal [3], but also in main group [4] and f-element coordination chemistry [5]. While it is presently almost impossible to survey the vast number of compounds of the cyclopentadienyl ligand known for virtually every element of the periodic table, the majority of transition metal complexes containing this ligand can be divided into three main classes. The first group comprises the structural analogues of ferrocene and are characterized by the parallel arrangement of both ring ligands. Apart from the ring-substituted ferrocene derivatives, the number of parallel metallocenes are rather limited and their study has been mainly motivated by the interest in the electronic, magnetic, and other physico-chemical properties [6]. The second class consists of bent metallocenes that contain two cyclopentadienyl ligands arranged in a non-parallel manner with additional coordination of up to two two-electron donors L or three one-electron ligands X [7]. Complexes of this type have gained enormous importance for the development of early transition metal chemistry, since their electronic structure has been thoroughly investigated and consistently established [8] and since the MCp_2 fragment is usually preserved as an organometallic template in most of the reactions. Finally the group of half-sandwich complexes having only one cyclopentadienyl ligand, but up to four L- and six X-type ligands is characterized by a large structural variety. The chemistry of these so-called piano-stool molecules is still developing rapidly and many interesting results are in the offing.

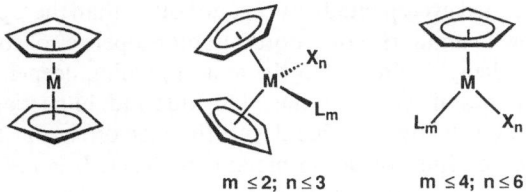

$m \leq 2;\ n \leq 3$ $\qquad$ $m \leq 4;\ n \leq 6$

The unique importance of the cyclopentadienyl ligand in the coordination chemistry of transition metals is mainly based on following three features:

1) The cyclopentadienyl ligand in the usual η^5-bonding mode formally occupies three coordination sites of a transition metal. The bonding is characterized by a relatively strong metal-ligand bonding interaction. Although quantitative thermodynamic data are still scarce [9], the dissociation of the cyclopentadienyl ligand seems to require much higher energy as compared to that for monodentate ligands such as carbon monoxide or substituted phosphine.

2) The cyclopentadienyl ligand is rather flexible in terms of its over-all electronic property. Thus both σ- and π-donating ability is normally prevailing, but electron-accepting interaction does play a certain role. The kinetic inertness of many cyclopentadienyl containing transition metal complexes is remarkably high, as impressively documented by the isolation of hydrolysis stable complexes [10], compounds having metal centers in their highest possible [11], or in very low oxidation states [12].

3) The geometry of the cyclopentadienyl-metal interaction is generally predictable to a high degree, since in most of the cases the η^5-bonding is encountered. When the cyclopentadienyl ligand is simplistically considered as a rotationally invariant monodentate ligand, it gives rise to familiar structures such as tetrahedra (e.g. MCp_2X_2 or $MCpL_3$) or octahedra (e.g. $MCpLX_4$). In certain complexes a fluxional behavior called ring slippage or haptotropic shift $(\eta^5-\eta^3-\eta^1)$ is observed, but this distortion does not account for the groundstate of the majority of cyclopentadienyl-metal complexes [13].

A further dimension was added to the variety of the cyclopentadienyl-metal complexes, when ring-substituted cyclopentadienyl ligands were introduced. While the use of methylcyclopentadienyl ligand Cp′ affects neither structure nor reactivity of the complexes profoundly, the systematic introduction of pentamethylcyclopentadienyl ligand Cp* resulted in the synthesis of a large number of novel complex types and in the discovery of new reactivity patterns not known for the corresponding Cp systems [14]. The reason for this difference was attributed on one hand to an enhanced donor property of the Cp* versus the Cp ligand resulting in a more covalent metal-ligand interaction and on the other hand, to the effective blocking of bimolecular reactions leading to lower molecular aggregates and/or coordinatively unsaturated species [15]. However, a clear-cut separation of these two influences has not been explicitly attempted so far, in contrast to the availability of even semi-quantitative data of ligand effects for phosphine ligands [16]. In general, there seems to be a dearth of systematic investigations into the complexation behavior of substituted cyclopentadienyl ligands other than the Cp* ligand.

In the present review I summarize the coordination properties of some sterically demanding cyclopentadienyl ligands with the aim of gaining deeper insights into the bonding characteristics of cyclopentadienyl ligands and, in particular, into the role of steric effects exerted by peripheral substituents on both structure and reactivity of the corresponding transition metal complexes. It is my belief that a more thorough understanding of the ligands' stereoelectronic influences as well as ligand-ligand interactions within the coordination sphere will provide the possibility of a more rational design of reactive transition metal complexes. In the context of synthetic applications tailoring metal complexes is becoming increasingly important, when specific properties such as those critical for stereoselective reagents or catalyst precursors are required.

The scope of the present paper is limited to those cyclopentadienyl ligands that contain more than two bulky substituents and transition metal complexes derived thereof in order to be able to focus on the specific effects of these ligand systems. A selection of some mono-substituted cyclopentadienyl ligands will be treated also. Among the numerous reviews highlighting special aspects of cyclopentadienyl metal complexes, there are some early reviews that are worth mentioning for historic reasons [17]. As more recent reviews an article on functionally substituted cyclopentadienyl ligands has appeared [18] and a comprehensive survey of half-sandwich complexes containing all monosubstituted cyclopentadienyl ligands is now available [19]. Another review thoroughly treats cyclopentadienyl ligands with electronegative substituents [20]. A new article on cyclopentadienyl metal halide complexes [21a] as well as a critical survey of synthetic procedures for cyclopentadienyl metal complexes have also been written [21b].

2 Alkyl-Substituted Cyclopentadienyl Ligands

Two commonly used synthetic methodologies for the synthesis of transition metal complexes with substituted cyclopentadienyl ligands are important. One is based on the functionalization at the ring periphery of Cp or Cp* metal complexes and the other consists of the "classical" reaction of a suitable substituted cyclopenta- dienyl anion equivalent and a transition metal halide or carbonyl complex. However, a third strategy of creating a specifically substituted cyclopentadienyl ligand from smaller carbon units such as alkylidynes and alkynes within the coordination sphere is emerging and will probably find wider application [22].

2.1 *tert*-Butyl-Substituted Cyclopentadienyl Ligands

2.1.1 Ligand Synthesis

The preparation of mono- and di-*tert*-butylcyclopentadienes 1 and 2 starting from monomeric cyclopentadiene was reported first in 1963 [23]. It was noted that the nucleophilic attack of the cyclopentadienide anion on *tert*-alkyl halide has to compete with elimination reaction giving isobutene. The yield of the di- and tri-*tert*-butylcyclopentadienes 2 and 3 was therefore reported to be modest to low [23, 24]. Recently an elegant improvement for this synthesis using phase transfer catalysis was presented (Eq. 1), but the availability of the tri-substituted derivative 3 is still hampered by the low yield [25]. It is important to note that for steric

R = H (4), CMe₃ (5)

reasons the *tert*-butyl groups in di-substituted cyclopentadienes **2** is always in the 1,3-position, whereas the tri-substituted derivative **3** is selectively obtained as the 1,3,5-isomer. It was moreover stated that placing three *tert*-butyl groups adjacent to each other on a cyclopentadiene ring is synthetically a very difficult task. 1,2,3,4-Tetra-*tert*-butylcyclopentadiene could be therefore prepared in only a minute amount in a tedious multi-step synthesis [26].

Mono- and 1,3-di-*tert*-butylcyclopentadienyllithium $Li(Bu_nCp)$ **4** ($n = 1$) and **5** ($n = 2$) can be obtained directly by addition of methyllithium to 6,6-dimethyl fulvene and 2-*tert*-butyl-6,6-dimethyl fulvene, respectively [27a]. Since the fulvenes themselves are easily accessible by base-induced condesation reactions [27b], this method offers a viable alternative to the usual metalation of the substituted cyclopentadienes **1** and **2** (Eq. 2).

2.1.2 (1,3-Di-*tert*-butylcyclopentadienyl)metal Complexes

In a report on the preparation of multiply *tert*-alkyl-substituted ferrocenes Friedel-Crafts alkylation of the parent ferrocene $FeCp_2$ with *tert*-butyl chloride was described [28]. Using various Lewis and Brønsted acids as catalysts up to two *tert*-butyl groups could be introduced into each ring (Eq. 3). Whereas the lower members of the *tert*-butylated ferrocenes were isolated as mixtures of positional isomers, orange tetra-*tert*-butylferrocene was shown to be exclusively the 1,1',3,3'-isomer **6**. This was also confirmed by an independent synthesis starting from $FeCl_2$ and 1,3-di-*tert*-butylcyclopentadienyllithium $Li(Bu_2Cp)$ (**5**) according to Eq. 4 [29] and by a single crystal X-ray structural analysis which revealed a conformation of the sandwich molecule determined by the interannular repulsions of the four bulky groups [30]. In 1981 it was discovered that $Fe(Bu_2Cp)_2$ exhibits

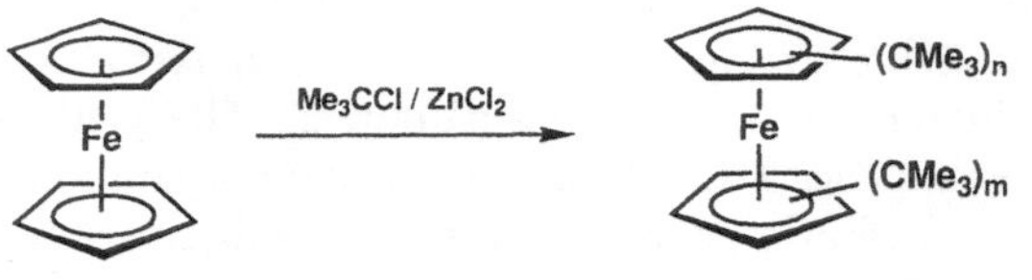

$$(3)$$

$$n + m \leq 4$$

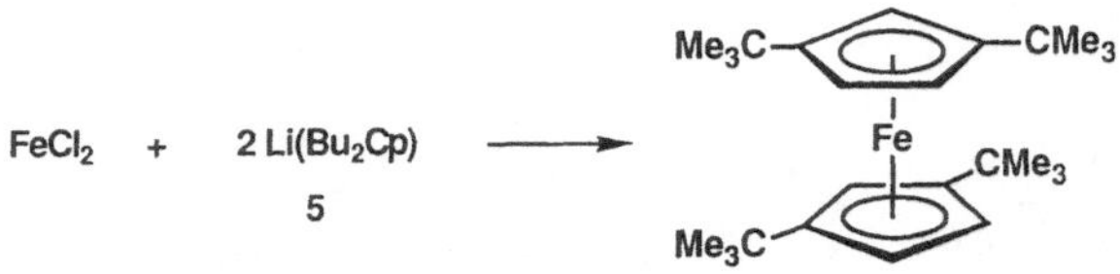

$$(4)$$

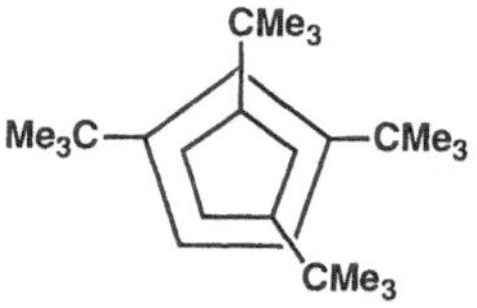

6A 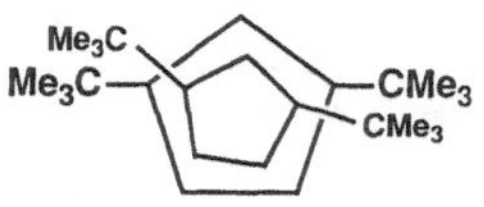   6B a hindered rotation of the five-membered rings about the metal-ring ligand axis in solution [31]. The barrier to ring rotation was high enough ($\Delta G^{\pm} = 13.1$ kcal mol^{-1}) to be determined for the first time using conventional variable-temperature NMR spectroscopy. Since enantiotopic groups become inequivalent at lower temperatures, the ground state is supposed to have only a C_2 axis, while an over-all C_{2v} or C_{2h} symmetry is observed above the coalescence temperature. This observation is in good agreement with the molecular structure of 6 in the solid state, where indeed a conformation with two different types of *tert*-butyl groups 6A was found. Consequently, the energetically most unfavorable transition state for the rotation 6B can be considered to be one in which all four ring substituents are eclipsed. Previously it had been postulated that upon protonation of Fe(Bu$_2$Cp)$_2$ with neat trifluroacetic acid, two rotational isomers are detectable by ^{1}H NMR spectroscopy [32]. As will be discussed in more detail later, the conformational rigidity due to hindered ring rotation seems to be one of the most prominent features of metallocenes containing sterically bulky cyclopentadienyl ligands.

A few half-sandwich iron complexes with the Bu$_2$Cp ligand are also known. The direct reaction of 2 with Fe$_2$(CO)$_9$ gives [Fe(Bu$_2$Cp)(CO)$_2$]$_2$ (7) which according to solution IR spectroscopic data predominantly exists as the *trans* isomer in the equilibrium mixture [32]. Mössbauer spectroscopic data are believed to show some effects of the bulky ring substituents. Reaction of the dimer with elemental sulfur and treatment of the resulting μ-sulfido bridged complex [Fe(Bu$_2$Cp)(CO)$_2$]$_2$(μ-S$_3$) (8) with aroyl chloride gives thiocarboxylate complexes Fe(Bu$_2$Cp)(CO)$_2$SCOR (9) (R = C$_6$H$_5$, 4-(NO$_2$)C$_6$H$_4$, 3,5-(NO$_2$)$_2$C$_6$H$_3$). The enhanced reactivity of the Bu$_2$Cp versus the parent Cp as well as BuCp dimer was attributed to the presence of a weaker Fe-Fe bond. The reaction of FeCl$_2$ with Li(Bu$_2$Cp) at low temperature yields under CO low yields of Fe(Bu$_2$Cp)-(CO)$_2$Cl (10), the main product being the ferrocene derivative 6 [35].

A remarkably rigid molecule in which the ring rotation is locked up to 100 °C was found in the form of Co(Bu$_2$Cp)(PMe$_3$)$_2$ (11) [36]. This complex was synthesized as a brown oil by electrophilic addition of *tert*-butyl bromide to Co(BuCp)(PMe$_3$)$_2$ followed by deprotonation of similarly rigid [Co(Bu$_2$Cp)-

(5)

103

$(PMe_3)_2H]^+$ (**12**) with NaH (Eq. 5) [37]. The conformation for these complexes as deduced from ^{31}P NMR spectroscopic data are depicted above. Interestingly the cationic derivative $[Co(Bu_2Cp)(PMe_3)_2Me]^+$ (**13**) possesses a molecular plane of symmetry.

By an analogous procedure related $[Co\{\eta^5\text{-}C_5H_3(i\text{-}Pr)_2\}(PMe_3)_2H]^+$ (**14**) and $Co[\eta^5\text{-}C_5H_3(i\text{-}Pr)_2](PMe_3)_2$ (**15**) were also synthesized, but were shown to contain a freely rotating ring ligand. This was ascribed to a gear-meshing mechanism during the ring rotation. Restricted rotation at lower temperatures was observed for $Co[\eta^5\text{-}C_5H_3(i\text{-}Pr)(t\text{-}Bu)](PMe_3)_2$ (**16**) and $Co[\eta^5\text{-}C_5H_3(i\text{-}Pr)(SiMe_3)](PMe_3)_2$ (**17**). Finally it was noted that all attempts at introducing a third bulky ring substituents by this method were unsuccessful.

The yellow cobaltocenium ion $[Co(Bu_2Cp)_2]^+$ (**18**), isoelectronic with the corresponding ferrocene $Fe(Bu_2Cp)_2$, has been prepared from $CoCl_2$ and $Li(Bu_2Cp)$ (Eq. 6) [35, 38] and the barrier to ring rotation was determined for the hexafluorophosphate of **18** by variable-temperature NMR spectroscopy ($\Delta G^{\ddagger} = 12.7 \text{ kcal mol}^{-1}$) [35]. The dark green nickelocene $Ni(Bu_2Cp)_2$ (**19**), synthesized analogously, was reported to be thermally remarkably robust [38].

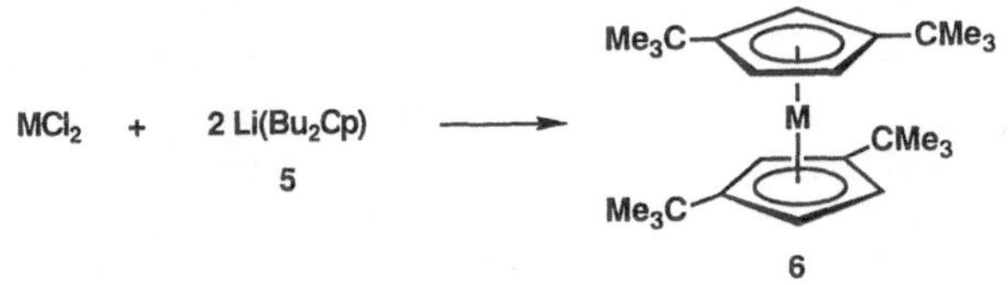

$$MCl_2 \quad + \quad 2 \text{ Li(Bu}_2\text{Cp)} \quad \longrightarrow \quad \tag{6}$$

5

6

M = Co, Co$^+$ (18),
Ni (19)

Reaction of $Li(Bu_2Cp)$ with $TiCl_4$ according to Eq. 7 gave moderate yields of the titanocene dichloride derivative $Ti(Bu_2Cp)_2Cl_2$ (**20a**) as orange crystals. Using potassium pentane-soluble blue Ti(III) complex $Ti(Bu_2Cp)_2Cl$ (**21a**) was obtained [39a] that can be also synthesized directly from $Li(Bu_2Cp)$ and $TiCl_3$ (Eq. 8) [35]. As has been observed previously for the preparation of decamethyltitanocene dichloride $TiCp^*_2Cl_2$ [40], coordination of substituted cyclopentadienyl anions at titanium centers invariably proceeds smoother and with higher yield when titanium(III) chloride is used instead of titanium(IV) chloride. The initially formed Ti(III) complex **21a** can be oxidized in situ to give the dichloride **20a** in good yield. The Ti(III) complex **21a** was shown by EPR, magnetic susceptibility measurement, and single crystal X-ray structure analysis to be monomeric with a temperature independent effective magnetic moment corresponding to one unpaired electron. In contrast, the chloride-bridged Cp complex $[TiCp_2(\mu\text{-}Cl)]_2$ shows an antiferromagnetic interaction [41]. This difference is accounted for by the steric repulsion between the two metallocene units, since the chloride bridge is known to be fairly weak. The structure of **21a** in the solid state revealed a conformation of the two Bu_2Cp ligands that is quite unusual in that the two pairs of *tert*-butyl groups are virtually eclipsed. Unexpectedly the dichloride derivative **20a** adopts a very similar conformation in the solid state. The steric congestion caused by this ligand coordination resulted in an elongation of the ring-titanium bond length, in particular ring-carbon atoms carrying the *tert*-butyl groups are farthest apart from the titanium center. An increased bending of the five-membered rings (121.0°) was also noted.

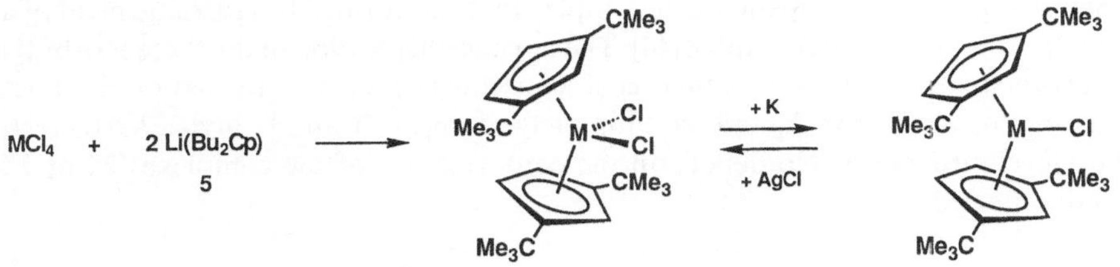

$$MCl_4 \ + \ 2\ Li(Bu_2Cp) \longrightarrow$$

5

M = Ti (a), Zr (b), Hf (c)

(7)   (8)

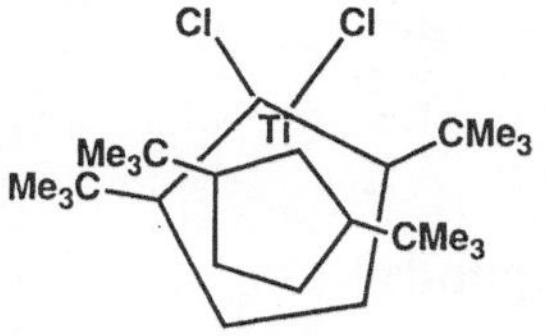

20a 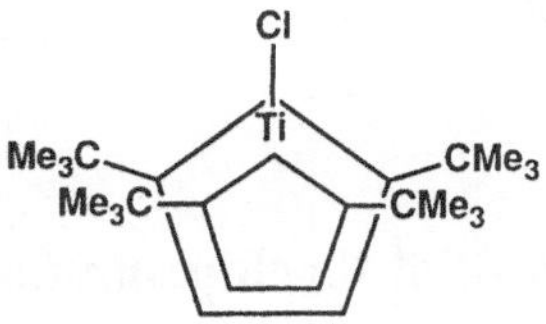   **21a**

The metallocene dichloride of zirconium and hafnium **20b** and **20c** were also prepared and underwent reduction with potassium to give *monomeric* metallocene monochloride complexes **21b** and **21c** (Eq. 8) [39b]. The structure of the zirconocene complex **21b** in the crystal showed a conformation which suggests a less steric strain as compared to **21a** due to zirconium's larger atomic size. As a consequence of the coordinative unsaturation an unusually short Zr−Cl bond length was found.

As half-sandwich complexes of titanium containing a Bu_2Cp ligand the orange crystalline trichloro complex $Ti(Bu_2Cp)Cl_3$ (**22**) [35] and red oily amido derivative $Ti(Bu_2Cp)(NMe_2)_3$ (**23**) are known. For **23**, ^1H and ^{13}C NMR spectroscopic data suggest a hindered rotation of the ring ligand relative to the titanium fragment [42].

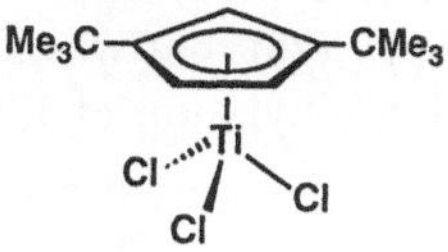

22 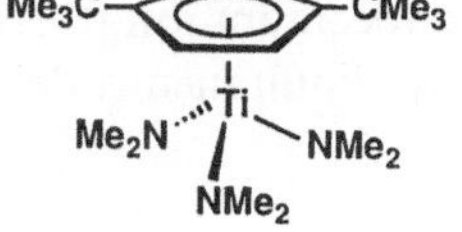   **23**

2.1.3 (Tri-*tert*-butylcyclopentadienyl)metal Complexes

Due to the difficulty in preparing the free ligand **3** complexes of the 1,2,4-tri-*tert*-butylcyclopentadienyl ligand are virtually unexplored. 1,2,4-tri-*tert*-butylcyclopentadienylsodium $Na(Bu_3Cp)$ was reacted with $Mo(CO)_6$ and the resulting anion $[Mo(Bu_3Cp)(CO)_3]^-$ treated with methyl iodide to give pale yellow $Mo(Bu_3Cp)(CO)_3CH_3$ (**24**) [43a], sofar the only transition metal complex with a Bu_3Cp ligand. An interesting rhodocenium cation derivative **25** containing 1,2,3-tri-*tert*-butylcyclopentadienyl ligand was formed by hydride abstraction of the corresponding cyclopentadiene complex that was formed by rearrangement of a vinylrhodacyclobutene complex [44]. The unusual disposition of the three *tert*-butyl substituents next to each other is a consequence of the formation because this cyclopentadienyl ligand is ultimately derived from 3-vinyl-1,2,3-tri-*tert*-butylcyclopropene-1. No details on the conformation of the complexes **24** or **25** were reported.

24   25

2.2 *iso*-Propyl-Substituted Cyclopentadienyl Ligands

Multiply *iso*-propyl-substituted cyclopentadienes can be prepared by repeated alkylation of the cyclopentadiene anion with *iso*-propyl bromide. However, the last alkylation step to give the penta-*iso*-propylcyclopentadiene proceeds only with low yield [24b]. Tetra-*iso*-propylcyclopentadienyl complexes 1,1′,2,2′,3,3′,4,4′-octa-*iso*-propylferrocene $Fe(Pr_4Cp)_2$ (**26**) [43b] and the molybdenum carbonyl complexes $Mo(Pr_4Cp)(CO)_3X$ (**27**) (X = H, CH_3, Cl, $SiMe_2H$) as well as the molybdenum-molybdenum bonded dimers $[Mo(Pr_4Cp)(CO)_2]_2$ and $Mo(Pr_4Cp)Cp^*(CO)_4$ [43a] were synthesized by conventional methods. The ^1H NMR signals of the ferrocene derivative **26** exhibits temperature dependence and this was interpreted as freezing out of the rotation about the metal-ligand axis. Three rotamers **26A–26C** were inferred at lower temperatures and the barrier to rotation was estimated to $\Delta G^{\pm}$ (15 °C) = 13.6 kcal mol^{-1}. At even lower temperatures the rotation of the *iso*-propyl substituents about the carbon-carbon bond is slowed down.

The first penta-*iso*-propylcyclopentadienyl complex reported was 1,2,3,4,5-penta-*iso*-propylcobaltocenium ion $[Co(Pr_5Cp)Cp]^+$ (**28**) that was elegantly prepared in high yield by simultaneous deprotonation/alkylation of 1,2,3,4,5-penta-

26   26A   26B   26C

methylcobaltocenium ion using excess $KOCMe_3$ and methyl iodide (Eq. 9) [45]. Analogous complex with five *iso*-pentenyl substituents $[Co\{\eta^5\text{-}C_5(3\text{-}C_5H_{11})_5Cp]^+$ (**29**) was prepared using similar methodology that had previously been developed for the synthesis of peralkyl-substituted arene ligands [46]. Both complexes show, at room temperature, a rigid chiral conformation of the peralkyl cyclopentadienyl ligand due to the gear-meshing of all five *iso*-alkyl groups in such a manner that all methine protons point in the same direction. This orientation causes a chirality that was first observed in hexa-substituted benzene derivatives and termed conformational directionality [47]. At higher temperatures, however, racemization of **28** and **29** occurs as depicted in Eq. 10 and from the temperature dependent NMR spectra the exchange of the methyl groups above and below the five-membered ring plane is clearly detected. An activation barrier of $\Delta G^{\neq}$ 17.1 and 19.4 kcal mol^{-1}, was determined for **28** and **29**, respectively. These values can be compared with that found for the racemization process of $Cr[\eta^6\text{-}C_6(SiMe_2H)_6](CO)_3$ ($\Delta G^{\neq} = 13.9$ kcal mol^{-1}) [47b]. The higher rigidity of **28** and **29** can be associated with the more flexible nature of the silyl groups in the latter complex, in agreement with the findings made for trimethylsilyl-substituted cyclopentadienyl ligands (vide infra).

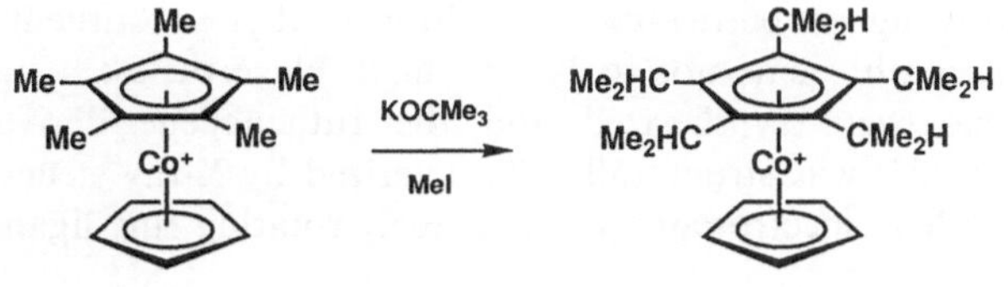

(9)

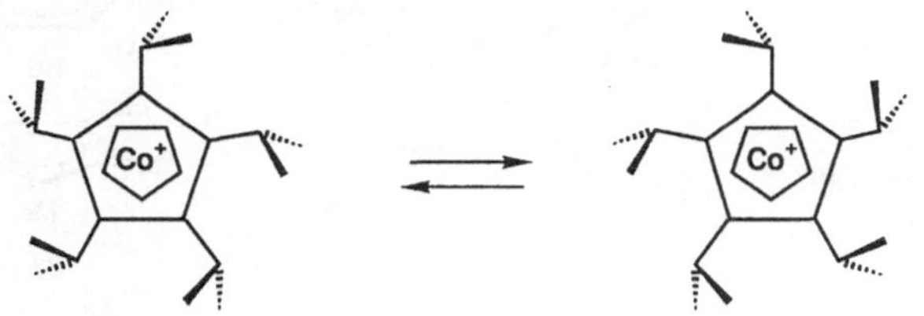

(10)

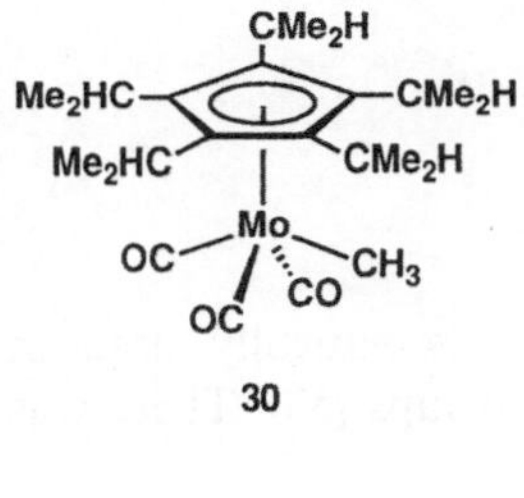

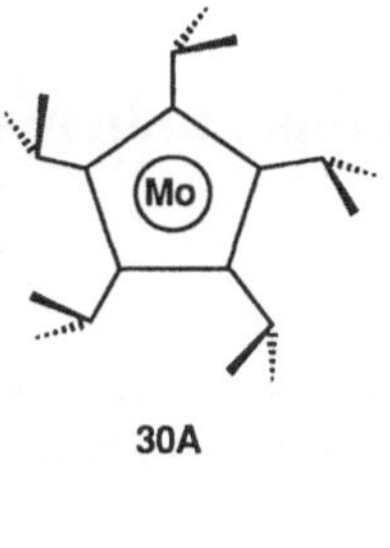

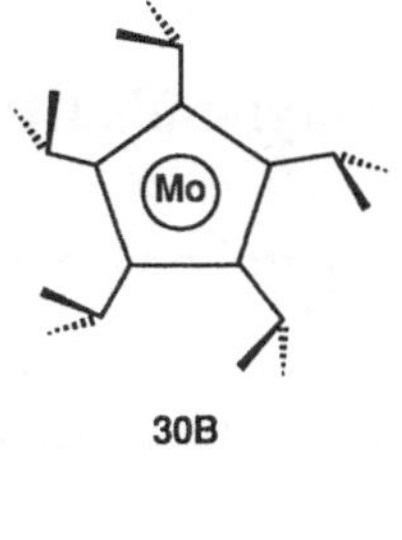

The only other $C_5(i\text{-}Pr)_5$ complex, $Mo(Pr_5Cp)(CO)_3CH_3$ (**30**) also shows fluxional *iso*-propyl groups at 100 °C, but below the coalescence temperature two isomers in the ratio 4:3 are obeserved according to ^{13}C NMR spectra. They are postulated to correspond to two chiral isomers one of which (**30A**) shows directional chirality, whereas in the second conformer **30B** one *iso*-propyl group is mismatched in its orientation [43a]. The origin of this irregularity is not known.

2.3 Other Alkyl-Substituted Cyclopentadienyl Ligands

In passing it should be mentioned that there are other bulky cyclopentadienyl ligands bearing five alkyl groups which upon complexation at transition metal centers should lead to potentially interesting complexes. Only a few pentabenzyl-cyclopentadienyl complexes such as decabenzylferrocene $Fe[\eta^5\text{-}C_5(CH_2Ph)_5]_2$ (**31**) [48a] and half-sandwich complexes of the type $M[\eta^5\text{-}C_5(CH_2Ph)_5](CO)_2$ (**32**) (M = Co, Rh) [49] as well as a gold complex $Au[\eta^1\text{-}C_5(CH_2Ph)_5](PPh_3)$ [48b] have been reported so far. Also, with an improved synthesis for pentaethylcyclopenta-diene [50] more pentaethylcyclopentadienyl complexes [22] should become available. In view of the fascinating stereochemical and conformational problems associated with transition metal complexes of hexaethylbenzene [51] complexes with an $\eta^5\text{-}C_5Et_5$ ligand also hold some promise.

Tetrakis(trifluoromethyl)cyclopentadienyl ligand $C_5(CF_3)_4R$ (R = H, $OSiEt_3$) was described recently [52], but as was expected on the ground of the extreme electron-withdrawing properties of the trifluoromethyl substituents, the metal-ligand bond strength seems not to be too high. Nevertheless, a few transition metal complexes were synthesized and the ruthenocene derivative $Ru[\eta^5\text{-}C_5(CF_3)_4H]Cp^*$ (**33**) was structurally characterized by X-ray structural analysis and shown by NMR spectroscopy to have freely rotating ring-ligands.

31   32   33

The prototype of an electron poor cyclopentadienyl ligand is the pentakis(carbo-methoxy)cyclopentadienyl ligand $C_5(CO_2Me)_5$ of which the coordination chemistry was extensively studied. A comprehensive review has just appeared in the literature [20].

3 Phenyl-Substituted Cyclopentadienyl Ligand

3.1 Ligand Synthesis

One of the first substituted cyclopentadienyl ligands systematically used in organotransition chemistry was those containing phenyl groups [53]. There was

already a substantial knowledge about phenyl-substituted cyclopentadienes existing [54] when ferrocene was discovered in 1951. Tetraphenylcyclopentadiene **34**
was first reported as early as in 1898 [54a] and pentaphenylcyclopentadiene **35**
was desribed in 1925 [54b]. Key intermediates are the readily available 2,3,4,5-
tetraphenylcyclopentadienone and the bromide **36**. A less symmetrical derivative
1-(4-*tert*-butylphenyl)-2,3,4,5-tetraphenylcyclopentadiene was prepared by a slight
modification, but so far used only to prepare Ge, Sn, and Pb complexes
[55]. Starting from tetraphenylcyclopentadienone tetraphenylcyclopentadiene
$C_5H_2Ph_4$, **34** is easily accessible by reduction with $LiAlH_4/AlCl_3$ in high yields
[56, 57]. Scheme 1 summarizes the synthetic procedures leading to tetra- and
pentaphenylcyclopentadiene **34** and **35**.

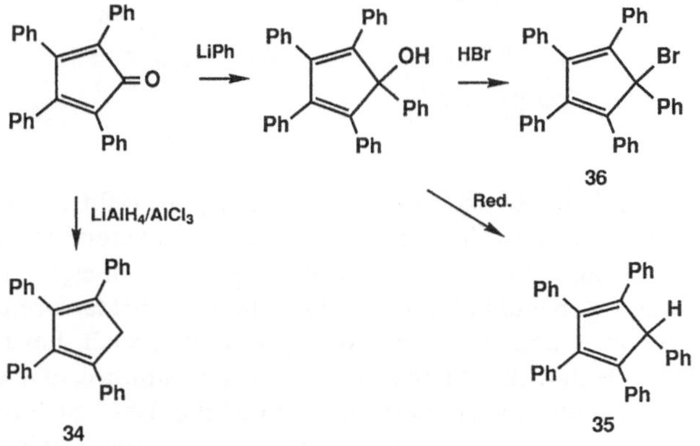

Scheme 1

As an early example for formation of a cyclopentadienyl ligand within the
coordination sphere of a transition metal 1,2,3,4-tetraphenylferrocene $Fe(\eta^5-C_5HPh_4)Cp$ was formed from diphenylacetylene and $FeCp(CO)_2CH_3$ [57]. Interestingly, in a number of transition metal $\eta^5-C_5Ph_5$ complexes the ligand were
generated from diphenylacetylene. These reactions must involve the scission of a
$C-C$ triple bond at a certain stage [58], bearing resemblance to the fundamental
step during alkyne metathesis. Multiple ring-arylation of cyclopentadienyl complexes, on the other hand, does not appear to have been performed to a wider extent.

3.2 (Tetraphenylcyclopentadienyl)metal Complexes

In view of some experimental difficulties associated with the pentaphenylcyclopentadienyl ligand the tetraphenylcyclopentadienyl ligand with one phenyl group less
has also been utilized. Metalation of tetraphenylcyclopentadiene **34** with
n-butyllithium in benzene cleanly gave $Li(C_5HPh_4)$ (**37a**) [59] and the potassium
derivative $K(C_5HPh_4)(THF)_{0.5}$ (**37b**) was prepared with KH in THF [60].

Following earlier reports [61] 1,1',2,2',3,3',4,4'-octaphenylferrocene $Fe(\eta^5-C_5HPh_4)_2$ (**38**) was straightforwardly obtained from $FeCl_2$ and $Li(C_5HPh_4)$ via nucleophilic substitution reaction according to Eq. 11 and isolated as brick-red crystals in good yield [59]. The smooth preparation of **38** appears to have some connection with the sterically less congested situation of the ligand sphere, since other ferrocenes bearing bulky ring substituents are not formed so easily. Electron spectroscopic and electrochemical data of **38** are consistent with a slightly electron-withdrawing property of the phenyl groups. On a preparative scale the one-electron oxidation with $AgPF_6$ afforded the corresponding blue ferrocenium cation $[Fe(\eta^5-C_5HPh_4)_2]PF_6$ (**39**).

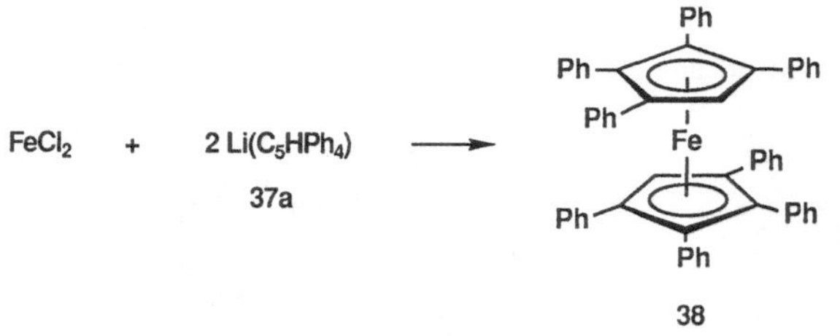

FeCl$_2$ + 2 Li(C$_5$HPh$_4$) $\longrightarrow$

37a

38

(11)

The single-crystal X-ray structural analysis of $Fe(\eta^5-C_5HPh_4)_2$ shows a parallel sandwich complex with a fairly long iron-ring distance of average 2.094 Å, although the iron-carbon distance for the unsubstituted cyclopentadienyl carbon is only 2.054 (3) Å, a value which is comparable to that found in other ferrocene derivatives. As expected the conformation of the two ring ligands is such that the hydrogen atoms of the cyclopentadienyl ligand are mutually positioned *anti* which conse-quently results in a fully staggered arrangement of the five-membered rings. The preference of this conformation (**38A**) over the other rotamers **38B** and **38C** can be explained by the interannular repulsions of the two $\eta^5-C_5HPh_4$ ligands with four pairs of phenyl groups avoiding each other. The phenyl substituents are arranged in a propeller-like manner and the dihedral angle the phenyl groups make with the cyclopentadienyl plane vary from 15.3 to 77.1°, whereby the two "inner" phenyl rings show greater deviation from coplanarity with the five-membered ring. In contrast to free tetraphenylcyclopentadiene [62] the phenyl rings cant in the same direction, whereas in the diene the rings are tilted in the direction of the unsubstituted carbon atom.

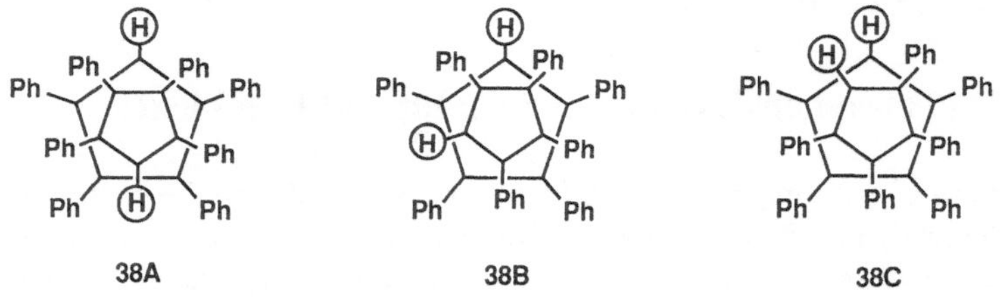

38A

38B

38C

Variable-temperature ^1H and ^{13}C NMR spectroscopic study of this seemingly crowded molecule **38** revealed that the ring rotation about the iron ring axis occurs freely down to −95 °C, although at lower temperatures the rotation of the inner phenyl groups about the *ipso*-carbon-cyclopentadienyl carbon bond becomes restricted. This is consistent with the conformation found in the crystal lattice and the free energy of activation was estimated to be ΔG^{*} ($-65\,^{\circ}C$) $= 9 \pm 1$ kcal mol^{-1}. The unexpected low barrier to the rotation of the η^5-C$_5$HPh$_4$ ligand can be accounted for by assuming a concerted gear-meshing mechanism to be operative.

1,1',2,2',3,3',4,4'-Octaphenylmetallocene of V, Cr, Co and Ni **40** were obtained by reacting the metal halide with K(C$_5$HPh$_4$)(THF)$_{0.5}$ (Eq. 12) and metallocenium cations of Cr, Co, and Ni were prepared by oxidation of the neutral metallocenes with AgPF$_6$ [60]. The paramagnetic metallocene complexes were all characterized by X-ray structural analysis and found to possess the same *anti* conformation as the ferrocene derivative **38** with a crystallographically imposed center of symmetry. The difference in metal-carbon bond lenghts between MCp$_2$ and M(η^5-C$_5$HPh$_4$)$_2$ complexes decreases on going from Fe to V, which suggests decreasing interactions between phenyl groups on opposing η^5-C$_5$HPh$_4$ rings. As the metal ring distances increase the conformation of the ligand increasingly resembles that found for the neutral tetraphenylcyclopentadiene.

$$MCl_2 \quad + \quad 2\,K(C_5HPh_4) \quad \longrightarrow \qquad\qquad\qquad (12)$$

37b

M = V, Cr, Co, Ni

40

Data from EPR spectroscopy and solid state magnetic susceptibility measurements are all comparable to those of the parent metallocenes MCp$_2$ and decamethylmetallocenes MCp$_2^*$ showing the same ground-state electronic configurations. The EPR measurements on Co(η^5-C$_5$HPh$_4$)$_2$ (**40c**) and [Cr(η^5-C$_5$HPh$_4$)$_2$]-PF$_6$ (**40b'**) suggested that the low symmetry of these complexes perturbs the metallocene energy levels more than Jahn-Teller distortions in the unsubstituted complexes. In general, the octaphenyl metallocenes showed much reduced reactivity in contrast to MCp$_2$ complexes.

In the context of pyridin synthesis from alkynes and nitriles homogeneously catalyzed by (cyclopentadienyl)cobalt complexes (Eq. 13), it was found that electron-withdrawing groups on the cyclopentadienyl ring significantly increase the activity [63]. During the screening of various cobalt half-sandwich complexes

$$2\,MeC\equiv CH \quad + \quad EtC\equiv N \quad \longrightarrow \qquad\qquad\qquad (13)$$

containing substituted cyclopentadienyl ligand, $Co(\eta^5\text{-}C_5HPh_4)(COD)$ (41) show-ed a high activity and both good chemo- and regioselectivity [63b]. Electronic effects of the ring substituents, expressed e.g. by the ^{59}Co chemical shift, correlate with various parameters for the catalytic reaction, but steric effects on the property of the catalytically active species, presumably to be the 14-electron half-sandwich fragment "$(C_5R_5)Co$" may be also inferred.

Green, monomeric 1,1',2,2',3,3',4,4'-octaphenyltitanocene chloride $Ti(\eta^5\text{-}C_5HPh_4)_2Cl$ (42) was prepared from $TiCl_3$ and $K(C_5HPh_4)$ [64]. While oxidation to the Ti(IV) complex with AgCl to give $Ti(\eta^5\text{-}C_5HPh_4)_2Cl_2$ (43) was successful (Eq. 14), attempts to prepare the titanocene $Ti(\eta^5\text{-}C_5HPh_4)_2$ remained in-conclusive. The monomeric structure of 42 was confirmed by EPR spectroscopy and by X-ray structural analysis. The conformation in the crystal is characterized by the two unsubstituted ring carbon atoms near each other. By tilting the two $\eta^5\text{-}C_5HPh_4$ ligands in such a way, the repulsive interaction between the two phenyl groups on opposing five-membered rings is minimized. Nevertheless the average titanium-ring distance of 2.085 Å is fairly long as is the bending angle of 136.4° comparatively small.

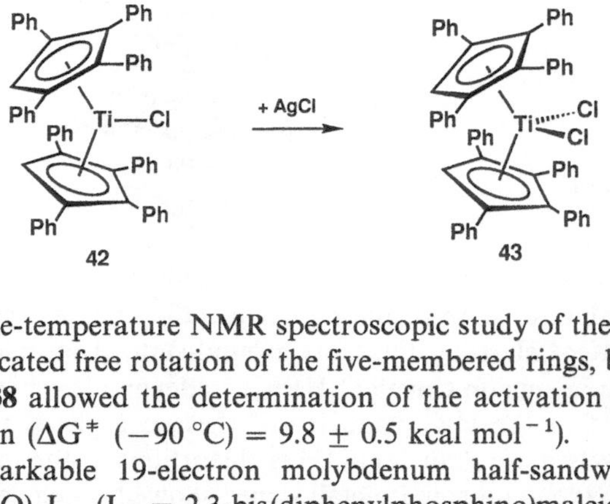

(14)

A variable-temperature NMR spectroscopic study of the titanium(IV) complex 43 also indicated free rotation of the five-membered rings, but, as in the ferrocene derivative 38 allowed the determination of the activation barrier for the phenyl ring rotation ($\Delta G^{\neq}$ ($-90\,°C$) = 9.8 ± 0.5 kcal mol^{-1}).

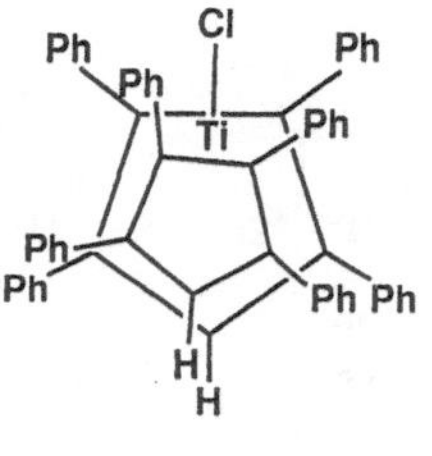

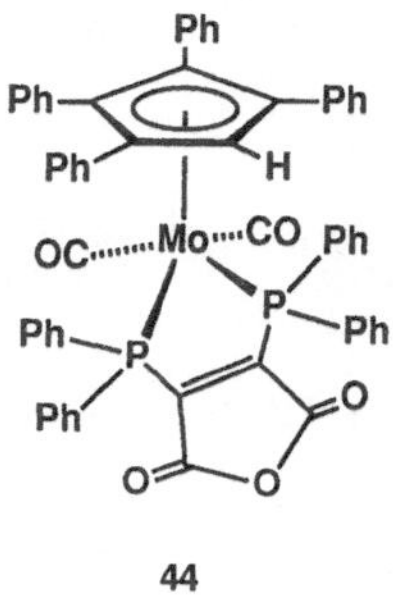

The remarkable 19-electron molybdenum half-sandwich complex $Mo(\eta^5\text{-}C_5HPh_4)(CO)_2L_2$ (L_2 = 2,3-bis(diphenylphosphino)maleic anhydride) (44) was prepared from $[Mo(\eta^5\text{-}C_5HPh_4)(CO)_3]_2$ and L_2 and its structure in the solid state determined [65]. The average distance from molybdenum to the ring ligand is within the normal range. The excess electron is localized in the chelating phosphine ligand and does not seem to affect the bonding to the five-membered ring significantly. The conformation of the η^5-C_5HPh_4 ligand is such that one ring phenyl group is in close proximity of a phenyl group at the phosphorus atom, whereas the other half of the L_2 ligand is situated below the unsubstituted carbon atom.

By using temperature dependent EPR spectroscopy the barrier to ring rotation was determined for **44**. Thus, at 31 °C two magnetically inequivalent phosphorus atoms are detected indicative of a rigid conformation such as found in the solid state. At 185 °C the two phosphorus atoms are equivalent because of fast rotation of the η^5-C_5HPh_4 ligand about the molybdenum ring axis. From the rate constants of this equilibration process at different temperatures the following activation parameters $\Delta H^{\neq} = 2.2 \pm 0.1$ kcal mol^{-1}; $\Delta S^{\neq} = -22.9$ cal K^{-1} mol^{-1} were obtained. The large negative entropy of activation was associated with a transition state of highly organized structure, since for the five-membered ring to rotate freely, the phenyl groups of the η^5-C_5HPh_4 ligand and the phenyl groups on the phosphorus atoms must rotate cooperatively in a gearing fashion. The alternative phosphorus-exchange pathway involving a trigonal-bipyramidal transition state was concluded to be unlikely because of unfavorable steric interactions between the phenyl groups of the η^5-C_5HPh_4 ligand and the phenyl groups on the phosphorus atoms.

1,2,3,4-Tetraphenylmolybdenocene dihydride Mo(η^5-C_5HPh_4)CpH$_2$ (**45**) was formed by addition of diphenylacetylene to MoCpL(PhC≡CPh)CH$_3$ (L = P(OMe)$_3$) (Eq. 15), presumably via an α-hydrogen abstraction to an intermediate methylidene hydrido complex, followed by addition of two equivalents of diphenylacetylene and C−H insertion with concomitant elimination of L [57b].

$$(15)$$

45

A structurally characterized gold complex with a C_5HPh_4 ligand Au(η^1-C_5HPh_4)(PPh$_3$) (**46a**) was reported [66]. An unusual trinuclear cluster **46b** derived therefrom was also characterized by X-ray structural analysis.

46a

46b

3.3 (Pentaphenylcyclopentadienyl)metal Complexes

Although pentaphenylcyclopentadiene has been known for so long, the number of transition metal complexes containing this polyarylated cyclopentadienyl ligand was limited until recently. This can be partly attributed to the low volatility and low solubility of the compounds in commonly used solvents, in addition to the rather uninformative ^{1}H and ^{13}C NMR spectra. There is also some limitation to the use of alkaline metal pentaphenylcyclopentadienyl compounds in the nucleophilic substitution reaction with metal halides because the formation of the persistent pentaphenylcyclopentadienyl radical [54b] often causes separation problems. It should be mentioned that main group complexes of the pentaphenylcyclopentadienyl ligand has also aroused considerable interest recently, in particular since the publication of the parallel sandwich structure of decaphenylstannocene $Sn(\eta^5\text{-}C_5Ph_5)_2$ with the rare S_{10} symmetry [67].

3.3.1 Complexes of Iron, Cobalt and Nickel

For the introduction of one pentaphenylcyclopentadienyl ligand the oxidative addition of bromopentaphenylcyclopentadiene 36 to homoleptic carbonyl complexes appears to be the most efficient way. Thus, reacting 36 with $Fe(CO)_5$ affords in good yields $Fe(\eta^5\text{-}C_5Ph_5)(CO)_2Br$ (47) isolated as dark red needles [61]. The molecular structure of 47 in the solid state was determined and shown to be that of the three-legged piano-stool type [61b]. The phenyl substituents are canted relative to the cyclopentadienyl ring at 49.1 and 142.8°, the iron-five-membered-ring distance being in the expected range. The half-sandwich complex 47 exhibits extensive reactivity as summarized in Scheme 2. The $\eta^5\text{-}C_5Ph_5$ ligand apparently imparts considerable kinetic inertness to complexes of which Cp (but not Cp*) analogues are known to be fairly labile. Thus, reaction with $NaBH_4$ gave thermally stable hydrido complex $Fe(\eta^5\text{-}C_5Ph_5)(CO)_2H$ (49) as yellow prisms decomposing at 279 °C, whereas reaction with sodium amalgam led to the isolation of the mercury-bridged complex $[Fe(\eta^5\text{-}C_5Ph_5)(CO)_2]_2(\mu\text{-}Hg)$ (50) [61]. Using $AlCl_3$ and benzene the arene cation could be prepared easily. Reduction of 47 to give $K[Fe(\eta^5\text{-}C_5Ph_5)(CO)_2]$ (51) can be accomplished using ultrasound-activated potassium. Treating the anion 51 with ethyl iodide to give the ethyl complex 52 followed by hydride abstraction led to the synthetically versatile ethylene complex $[Fe(\eta^5\text{-}C_5Ph_5)(CO)_2(\eta^2\text{-}C_2H_4)]^+$ (53) [69].

The ruthenium analogue of 47 $Ru(\eta^5\text{-}C_5Ph_5)(CO)_2Br$ (48) is also available, when $Fe(CO)_5$ is replaced by $Ru_3(CO)_{12}$ [68]. A wide range of substitution products were obtained through replacement of both carbonyl and bromide ligand against two-electron ligands L such as phosphines, phosphites, and ethylene. Electrochemistry of these derivatives were studied in some detail.

Although 1,2,3,4,5-pentaphenylferrocene $Fe(\eta^5\text{-}C_5Ph_5)Cp$ (54) was cleanly synthesized by thermal decarbonylation of $Fe(\eta^5\text{-}C_5Ph_5)(CO)_2(\eta^1\text{-}C_5H_5)$ (55) which had been obtained by reacting the bromide 47 with NaCp. Analogous methodology using $Na(C_5Ph_5)$ led to a green unidentified solid. In the initial work [61] it was therefore stated that all attempts at preparing decaphenylferrocene $Fe(\eta^5\text{-}C_5Ph_5)_2$

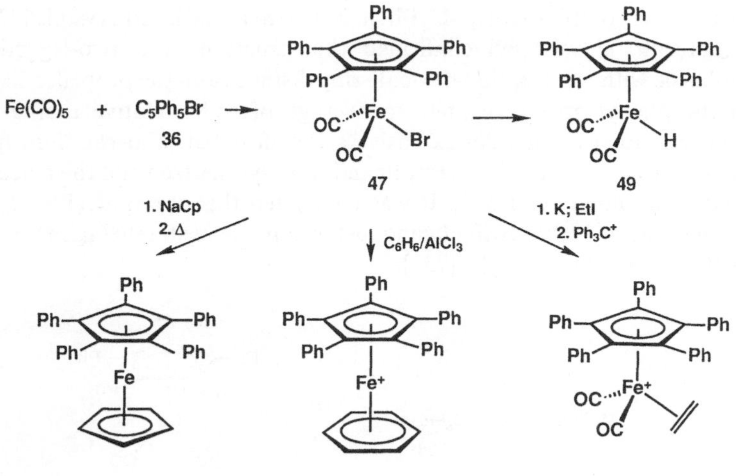

$$Fe(CO)_5 \;+\; C_5Ph_5Br \;\longrightarrow$$
$$\qquad\qquad\quad 36$$

Scheme 2

failed due to steric hindrance. Later, the preparation of pale yellow $Fe(\eta^5\text{-}C_5Ph_5)_2$ was claimed to have been achieved in 40% yield under somewhat different conditions, but neither experimental details nor any characterizing data were given [70].

More recently, an interesting linkage isomer of decaphenylferrocene **57** was synthesized by refluxing C_5Ph_5Br, zinc, and $Fe(CO)_5$ in benzene (Eq. 16) [71]. The intensely blue crystals of **57** was assigned the structure of an unusual zwitterion on the basis of spectroscopic and electrochemical data in which the $Fe(\eta^5\text{-}C_5Ph_5)$ fragment is coordinated to one of the five phenyl groups of the second C_5Ph_5 ligand in an η^6-fashion. In accordance with this unusual structure protonation with HBF_4 affords the red cation $[Fe(\eta^5\text{-}C_5Ph_5)(\eta^6\text{-}C_6H_5)C_5HPh_4]^+$ (**57a**). The NMR spectra of **57a** indicate that this cation no longer has the symmetry of the zwitterion, but that the $(\eta^6\text{-}C_6H_5)C_5HPh_4$ ligand is locked in a skewed conformation where all carbon atoms of the cyclopentadiene ring are non-equivalent.

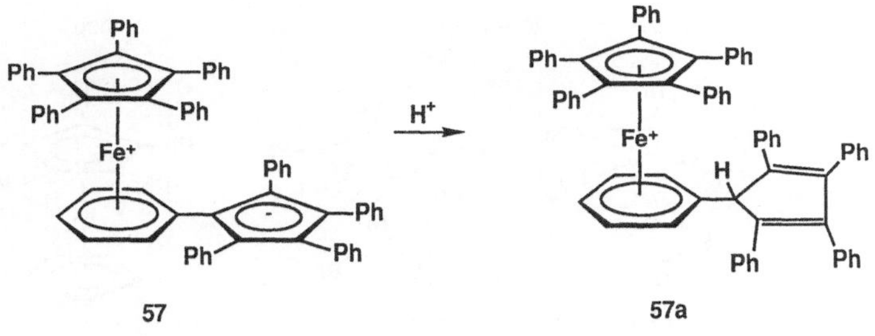

$$(16)$$

Reaction of the bromide **36** with tetracarbonylcobaltate $[Co(CO)_4]^-$ [72] or more conveniently with $Co_2(CO)_8$ in the presence of zinc (Eq. 17) gives the dicarbonyl complex $Co(\eta^5\text{-}C_5Ph_5)(CO)_2$ **(58)** as dark red crystals [73a]. An X-ray structural analysis [49] confirmed the structure of a two-legged piano-stool molecule with unexceptional cobalt-ring distance and the propeller-like orientation of the phenyl groups, canted an average of 55.8° relative to the plane of the cyclopentadienyl ring. An extensive series of cobalt and rhodium half-sandwich complexes bearing an $\eta^5\text{-}C_5Ph_5$ ligand were synthesized and their electrochemistry studied in much detail [73]. It was concluded that the $\eta^5\text{-}C_5Ph_5$ ligand behaves as an electron-withdrawing ligand, being capable of stabilizing radical anions such as $[Co(\eta^5\text{-}C_5Ph_5)(CO)_2]^-$ [73c].

$$Co_2(CO)_8 \ + \ \underset{36}{C_5Ph_5Br} \ \xrightarrow{\ Zn\ } \ \underset{58}{\text{(Ph-ring-Co-(OC)(CO))}} \qquad (17)$$

Oxidative addition of the bromide **36** to $Ni(CO)_4$ affords labile $Ni(\eta^5\text{-}C_5Ph_5)(CO)Br$ **(59)** that loses carbon monoxide easily to give a thermally stable dimer of the type $[Ni(\eta^5\text{-}C_5Ph_5)Br]_2$, **(60)** of which neither the Cp [74a] nor the Cp* [74b] analogue has so far been characterized completely. Complexes **59** and **60** represent a very versatile starting material for the synthesis of half-sandwich complexes bearing the $Ni(\eta^5\text{-}C_5Ph_5)$ fragment (Scheme 3). The carbonyl ligand can be thermally displaced by a donor ligand L such as phosphite and phosphine to give complexes of the type $[Ni(\eta^5\text{-}C_5Ph_5)(L)Br]$ **(62)**. Both the carbonyl and bromide ligand can be easily removed by $AgPF_6$ or $TlBF_4$ leaving an equivalent for the hypothetical cation $[Ni(\eta^5\text{-}C_5Ph_5)]^+$ as its bis(acetonitrile) complex $[Ni(\eta^5\text{-}C_5Ph_5)(CH_3CN)_2]^+$. Addition of tetramethylthiuram disulfide gave the structurally characterized 17-electron nickel(III) dithiocarbamate complex $[Ni(\eta^5\text{-}C_5Ph_5)(\eta^2\text{-}S_2CNMe_2)]^+$ **(61)** [76].

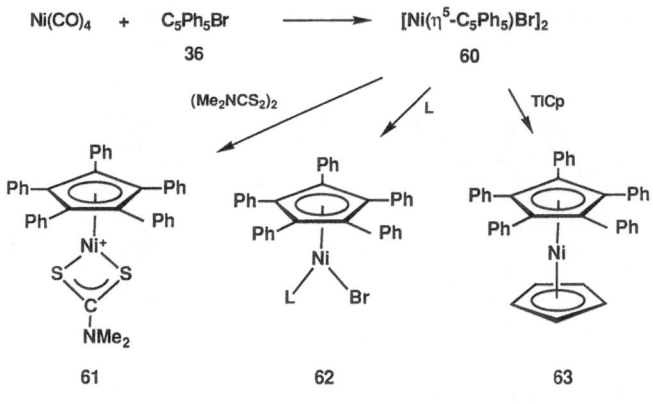

Scheme 3

When dimeric **60** is treated with TlCp olive green 1,2,3,4,5-pentaphenyl-nickelocene Ni(η^5-C$_5$Ph$_5$)Cp **(63)** was formed [72]. As expected **63** exhibits magnetic moments both in solution and in the crystal consistent with two unpaired electrons. Cyclic voltammetry showed that similar to decamethylnickelocene NiCp$_2^*$ [76] Ni(η^5-C$_5$Ph$_5$)Cp can be reversibly oxidized to the mono- and dication. This was ascribed to the steric bulk of the η^5-C$_5$Ph$_5$ ligand preventing rapid ligand loss from the oxidized species.

By reacting NiL$_3$ (L$_3$ = *E,E,E*-1,5,9-cyclododecatriene) with the persistent radical C$_5$Ph$_5$ orange diamagnetic decaphenylnickelocene Ni(η^5-C$_5$Ph$_5$)$_2$ **(64)** was obtained [77]. In the absence of structural details, however, it is not clear why this nickelocene derivative has paired electrons and whether this complex in fact contains two η^5-C$_5$Ph$_5$ ligands. In another peculiar transformation NiBr$_2$ gave with the dianion of the 1,2,3,4,5-pentaphenylaluminol bis(tetraphenylcyclobuta-diene)nickel **65** and a peculiar dinuclear complex **66** that contains a Ni(η^5-C$_5$Ph$_5$) fragment bridged to a Ni(η^4-C$_4$Ph$_4$) moiety by a 1,2,3-triphenylallenylidene ligand [78].

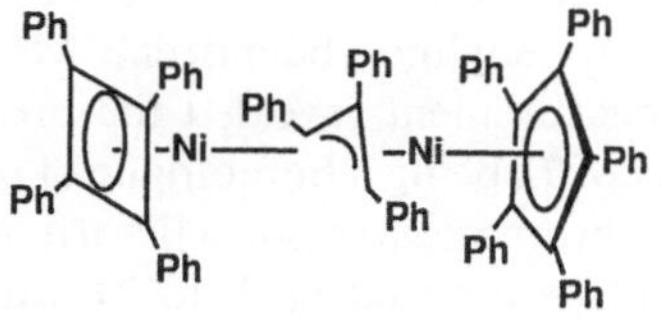

66

3.3.2 Complexes of Palladium

In a remarkable reaction according to Eq. 18 palladium(II) acetate reacts in methanol with diphenylacetylene to form dinuclear [Pd(η^5-C$_5$Ph$_5$)]$_2$(μ-PhC$\equiv$CPh) **(67)** [79]. The reaction mechanism was studied in some detail and part of the alkyne molecule was found as benzoic acid orthoester PhC(OMe)$_3$. When

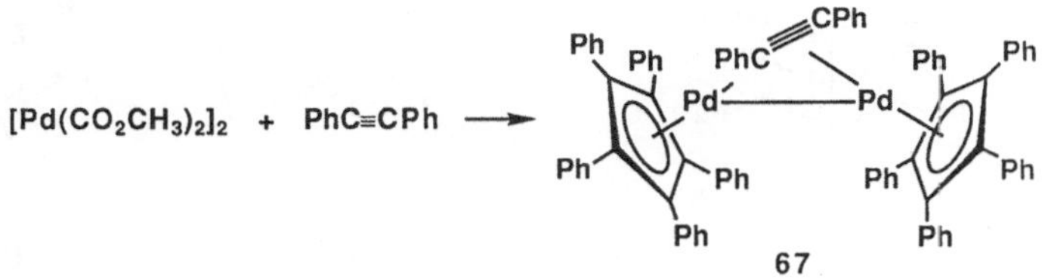

[Pd(CO$_2$CH$_3$)$_2$]$_2$ + PhC≡CPh $\longrightarrow$

67

(18)

[Pd(allyl)Cl]$_2$ + Na(C$_5$Ph$_5$) $\longrightarrow$

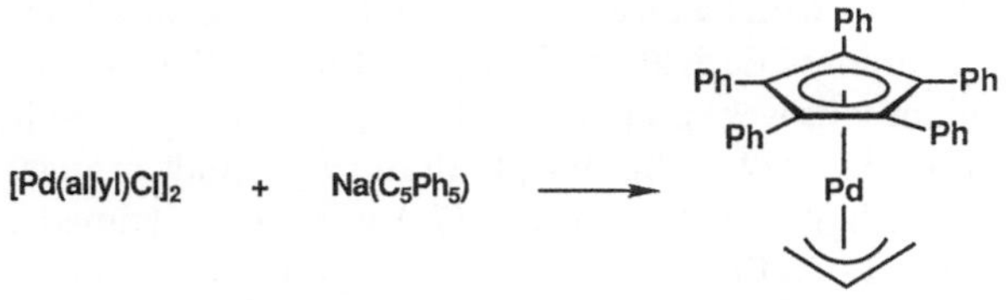

68

(19)

J. Okuda

exo-methoxy triphenylcyclobutenyl palladium derivative was reacted with various alkynes $RC \equiv CR'$ a similar type of complex $[Pd(\eta^5-C_5Ph_3RR')]_2(\mu-RC \equiv CR')$ was obtained [79a]. In addition to an X-ray structural analysis [79b] electrochemical studies were performed on **67** [79c].

When $Na(C_5Ph_5)$ was reacted with dimeric allyl palladium chloride (Eq. 19) a series of half-sandwich complexes of the general type $Pd(\eta^5-C_5Ph_5)(allyl)$ **(68)** could be synthesized [80]. Remarkably $P(OPh)_3$ induced a reductive elimination to give $Pd[P(OPh)_3]_4$ and allylpentaphenylcyclopentadiene. In the case of 2-chloroallyl derivative thermolysis gave dimeric $[Pd(\eta^5-C_5Ph_5)Cl]_2$ **(69)** that was also obtained by reacting $PdCl_2(PhCN)_2$ with $Na(C_5Ph_5)$. As in the case of the nickel complexes **60** described above this presumably chloride-bridged complex reacted with various two-electron donors such as CO to give mononuclear complex of the formula $Pd(\eta^5-C_5Ph_5)(L)Cl$ **(70)**, whereas reduction with zinc in the presence of diphenylacetylene afforded dimeric **67**.

3.3.3 Complexes of Early Transition Metals

Thermolysis of diphenylacetylene and $Mo(CO)_6$ gave in low yield a red sparingly soluble solid that was described as decaphenylmolybdenocene $Mo(\eta^5-C_5Ph_5)_2$ **(71)** [81]. The magnetic susceptibility measurement revealed the presence of two unpaired electron consistent with the fact of **71** being a homologue of paramagnetic chromocene $CrCp_2$. Oxidation of **71** with bromine gave green $[Mo(\eta^5-C_5Ph_5)_2]Br_3$ with three unpaired electrons that could be reduced back to **71** with Mg. Since the parent molybdenocene $MoCp_2$ has been characterized as a highly reactive transient with a triplet ground state undergoing rapid bimolecular reactions [82], giving e.g. dinuclear species such as $[MoCp(\mu-\eta^1:\eta^5-C_5H_4)]_2$, the isolation of **71** can be considered as a consequence of steric protection of a coordinatively unsaturated species.

71   72   73
            M = Cr, Mo, W   M = Ti, Zr

By heating $Li(C_5Ph_5)$ with hexacarbonyl of molybdenum and tungsten $M(CO)_6$ (M = Mo, W) the anionic derivatives $[M(\eta^5-C_5Ph_5)(CO)_3]^-$ as their lithium or tetraethylammonium salts were prepared and converted into the methyl derivatives $M(\eta^5-C_5Ph_5)(CO)_3CH_3$ **(72)** [70]. Purple titanocene and orange zirconocene dichloride $M(\eta^5-C_5Ph_5)Cl_2$ (M = Ti, Zr) **(73)** were also claimed to have been prepared by reacting $Li(C_5Ph_5)$ with the corresponding metal tetrahalide, but no details have been disclosed so far [70].

4 Silyl-Substituted Cyclopentadienyl Ligands

Trimethylsilyl-substituted cyclopentadienyl ligands Si_nCp (n = 1, 2, 3) have been extensively used in main group [4] and f-element coordination chemistry [5], whereas until recently only a few reports on transition metal complexes of these ligands other than for n = 1 were published. The use of 1,3-bis(trimethylsilyl) and 1,2,4-tris(trimethylsilyl)cyclopentadienyl ligands Si_2Cp and Si_3Cp has enabled the synthesis of an extensive series of novel complex types for main group elements [4], lanthanides, and actinides [83]. In these cases the increased covalent character of the metal-ligand bond in addition to the steric bulk exerted by the trimethylsilyl groups on the cyclopentadienyl ring appear to be the crucial effects, since the nature of metal-ligand bond for s, p, and f-elements is mainly influenced by the ratio of charge and size. Enhanced solubility in unpolar organic solvents and high crystallinity have also assisted the remarkable progress in this field.

4.1 Ligand Synthesis and Properties

The preparation of cyclopentadienes with up to four trimethylsilyl groups can be performed easily on a large scale starting with monomeric cyclopentadiene by repeated metalation with n-butyllithium and treating the resulting anion with chlorotrimethylsilane [84]. Any complication caused by formation of regioisomers does not occur, since all trimethylsilyl-substituted cyclopentadienes are fluxional by virtue of proto- and silatropic shifts [85]. Upon deprotonation with n-butyllithium the thermodynamically most favorable anion is formed selectively (Eqs. 20, 21). Thus, metalation of bis(trimethylsilyl)cyclopentadiene 74, which exists preferentially as the 5,5-isomer, selectively affords the 1,3-substituted anion 75. Similarly, tris(trimethylsilyl)cyclopentadiene 76, which is found to be mainly as the 2,5,5-isomer, affords the 1,2,4-substituted anion 77.

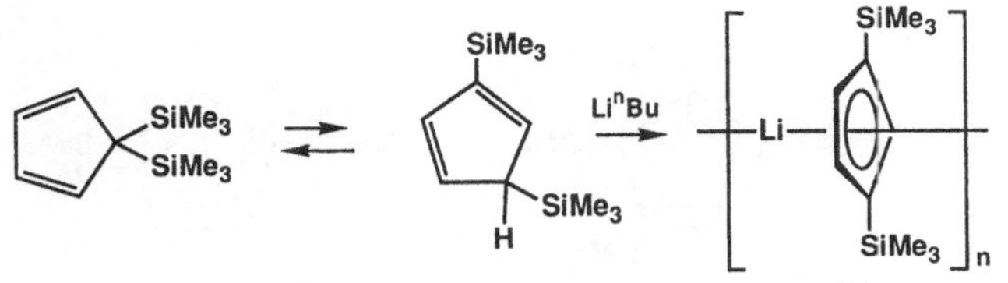

$$(20)$$

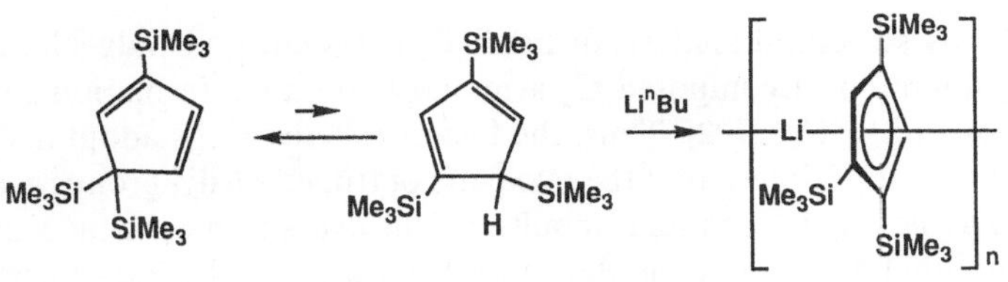

$$(21)$$

From the general substituent effect of a trimethylsilyl group (weak σ-donor, weak π-acceptor [86]) it can be expected that trimethylsilyl substituents on cyclopentadienyl ligands do not act as strongly electron-donating as alkyl-substituents do. This has been indeed verified in several instances by electro-chemical and various spectroscopic methods [87–89]. Results for a series of highly substituted ferrocene derivatives for instance have shown that trimethylsilyl groups on the cyclopentadienyl ligand can be characterized as weakly electron-with-drawing, but that the effects are fairly small [89]. Consequently, any difference between the coordination chemistry of Cp or Cp* and multiply trimethylsilyl-substituted cyclopentadienyl ligands may be primarily attributed to steric in-fluences.

4.2 [1,3-Bis(trimethylsilylcyclopentadienyl)]metal Complexes

4.2.1 Complexes of Iron, Cobalt, and Nickel

Although mono- and 1,1'-bis(trimethylsilyl)ferrocene were synthesized as early as in 1960 starting from ferrocene [90], it was not until 1978 that 1,1',3,3'-tetrakis(trimethylsilyl)ferrocene $Fe(Si_2Cp)_2$ (**78**) was reported [91]. The reaction of $Li(Si_2Cp)$ (**75**) with ferrous chloride according to Eq. 22 proceeds smoothly and affords **78** as orange air- and moisture-stable crystals in good yields [91, 92]. Evidence for a stepwise coordination of the two Si_2Cp ligands via a half-sandwich intermediate came from the isolation of a small amount of $Fe(Si_2Cp)(CO)_2Cl$ (**79**), when the complexation reaction was performed at low temperatures under CO [92].

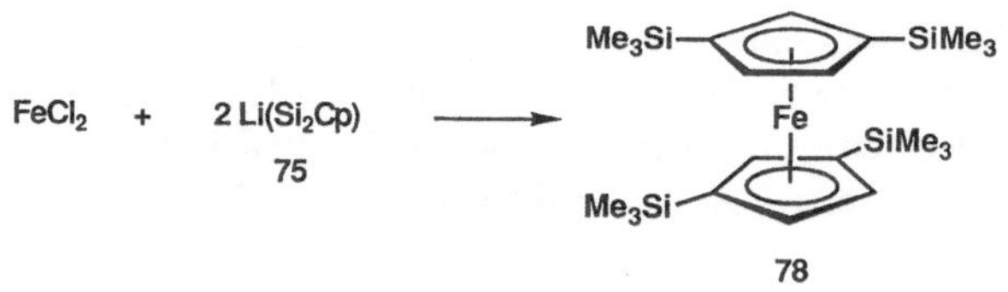

$$(22)$$

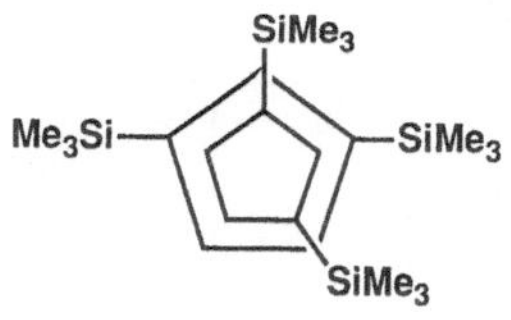

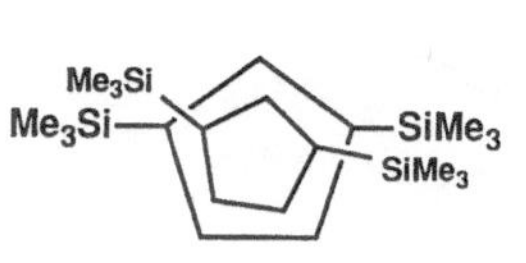

An X-ray structural analysis of $Fe(Si_2Cp)_2$ revealed a sandwich molecule with a crystallographically imposed C_2 symmetry and a conformation similar to that found for $Fe(Bu_2Cp)_2$ [92]. Thus, the five-membered rings adopt a virtually fully eclipsed conformation so that the two pairs of trimethylsilyl groups are interlocked and can avoid any interannular repulsion. The average metal-ring distance is fairly normal with 1.660 Å and the deviation from a parallel arrangement of the two rings is not large.

Variable-temperature ^{1}H NMR spectroscopic data of **78** indicated that the rotation of the Si_2Cp ligand about the iron-ring vector is restricted at temperatures below $-50\,°C$. The free energy of activation for rotation can be estimated to be $\Delta G^{\ddagger} = 11.0 \pm 0.5$ kcal mol^{-1}. A qualitative energy profile is shown in Fig. 1 where all possible 10 rotamers are placed according to their approximate energy contents as deduced on the simplistic assumption that interannular repulsion of the trimethylsilyl groups in a rigid sandwich molecule is solely responsible for a specific conformation. [92]. Each rotamer is designated whether its five-membered rings are eclipsed (e) or staggered (s). The sum of repulsions is given by a subscript counting staggered interaction of two trimethylsilyl groups in different rings as 0.25 and a full staggered one as 0.5. The ground state e_0 (or its enantiomer e'_0) obviously corresponds to the solid state structure with the four substituents in a fully staggered arrangement, also in agreement with the low-temperature limiting spectrum suggesting C_2 symmetry. Rotamers e_2 and s_1 are of higher energy and their population at temperatures above the coalescence temperatures would explain the apparent C_{2v} or C_{2h} symmetry as deduced from NMR spectra. The barrier to rotation can then be related to the energy difference $E(e_2) - E(e_0)$. The slightly lower value for $Fe(Si_2Cp)_2$ as compared to that of $Fe(Bu_2Cp)_2$ can be explained by the assumption that Bu_2Cp ligand is more rigid due to a shorter bond between the ring carbon and the carbon atom of the *tert*-butyl group.

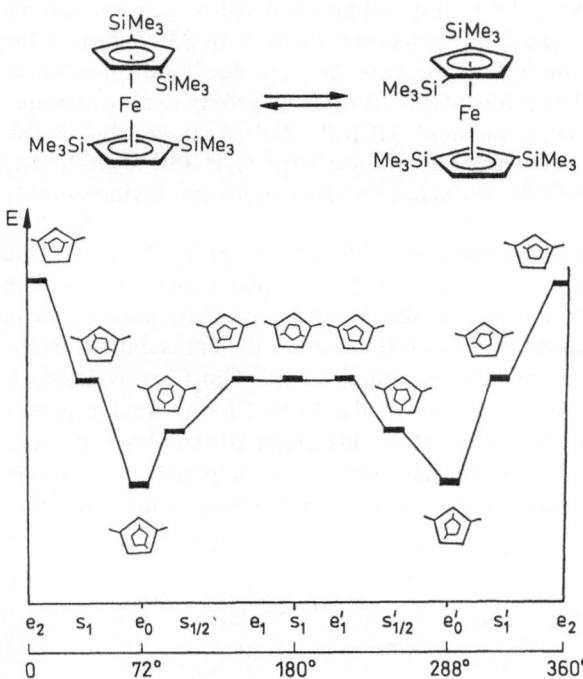

Fig. 1. Rotational profile for 1,1′,3,3′-tetrakis(trimethylsilyl)ferrocene **78**, as deduced from simple molecular model consideration. For the designation of the various rotamers see text

1,1′,3,3′-Tetrakis(trimethylsilyl)metallocene of cobalt and nickel $M(Si_2Cp)_2$ (**80**) (M = Co [93], Ni [91]) were prepared in an analogous way as described for the ferrocene derivative **78** (Eq. 23) and isolated as thermally robust black and green crystals, respectively. The oxidation of the cobaltocene with aqueous Fe^{3+} resulted in the formation of the yellow cobaltocenium ion $[Co(Si_2Cp)_2]^+$ (**81**) isolated as its hexafluorophosphate. By variable-temperature ^1H NMR spectroscopy a rotational barrier of 11.0 ± 0.5 kcal mol^{-1} was determined for **81**. Again the ΔG^+ value is slightly lower than that found for $[Co(Bu_2Cp)_2]^+$. As a catalyst precursor for the cobalt-catalyzed pyridine synthesis $Co(Si_2Cp)(COD)$ (**82**) was reported to show a relatively high activity [63b].

$$MCl_2 \quad + \quad 2\,Li(Si_2Cp) \quad \longrightarrow \quad \text{(structure)} \tag{23}$$

75

M = Co (80), Co⁺ (81), Ni

4.2.2 Complexes of Early Transition Metals

The half-sandwich complex $Ti(Si_2Cp)Cl_3$ (**83**) was prepared by reacting tris(trimethylsilyl)cyclopentadiene **74** with $TiCl_4$ (Eq. 24) and isolated as pentane-soluble orange-yellow crystals [93b]. The X-ray structural analysis of **83** confirmed the piano-stool configuration in the solid state with the expected 1,3-disposition of the trimethylsilyl groups on the cyclopentadienyl ring and geometrical parameters similar to those of the parent compound $TiCpCl_3$ [94]. Analogously, similar trichloro titanium complexes of the general formulae $Ti[\eta^5-C_5H_3(SiMe_3)R-1,3]Cl_3$ (**84**) (R = CMe_3 [95], BBr_2, $B(OEt)_2$ [96b]) have been obtained employing suitably substituted cyclopentadienes.

Both zirconium and hafnium homologues of **83** $M(Si_2Cp)Cl_3$ (M = Zr (**85**), Hf) were synthesized using the same methodology and isolated as white powders [94]. It was noted that unlike the mono- and bis(trimethylsilyl)cyclopentadiene tris(trimethylsilyl)cyclopentadiene **76** selectively reacts to the half-sandwich compounds and no formation of the metallocene complex $M(Si_2Cp)_2Cl_2$ was detected. From the insolubility an oligomeric nature similar to that found for the parent compound $MCpCl_3$ was suggested, although freshly prepared samples were more soluble. The zirconium derivative **85** has also been prepared by reacting $Mg(Si_2Cp)_2$ with $ZrCl_4$ and the crystals were reported to be triboluminescent [97].

$$MCl_4 \quad + \quad \text{(structure)} \quad \longrightarrow \quad \text{(structure)} \tag{24}$$

76

M = Ti (83), Zr (85), Hf

The metallocene dichloride complexes of the type $M(Si_2Cp)_2Cl_2$ (M = Ti (**86**) [91], Zr (**90**) [91, 98], Hf [91]) can be all synthesized by the nucleophilic substitution of the metal tetrachloride with $Li(Si_2Cp)$ (Eq. 25). For the titanocene derivative $Ti(Si_2Cp)_2Cl_2$ (**86**) the reaction of $Li(Si_2Cp)$ [99] or $Mg(Si_2Cp)_2$ [100] with titanium trichloride was found to give the isolable Ti(III) complex $Ti(Si_2Cp)_2Cl$ (**87**) that can be conveniently oxidized to the titanocene dichloride complex **86** (Eq. 26). Pentane-soluble, purple monochloride **87** was shown by mass spectrometry and EPR spectroscopy to be monomeric. The X-ray structure determination [100] confirmed this formulation with a Ti−C(ring) distance of average 2.368 Å, reflecting some steric encumbrance around the titanium center. The conformation is such that both pairs of the trimethylsilyl groups are more turned away from the open side of the metallocene unit as was found for the *tert*-butyl groups in $Ti(Bu_2Cp)_2Cl$ (**21**). Reaction of **87** with $NaBH_4$ gave $Ti(Si_2Cp)_2BH_4$ (**88**) which is also a monomer and proposed to have an η^2-bonded tetrahydridoborate ligand on the basis of IR data [100].

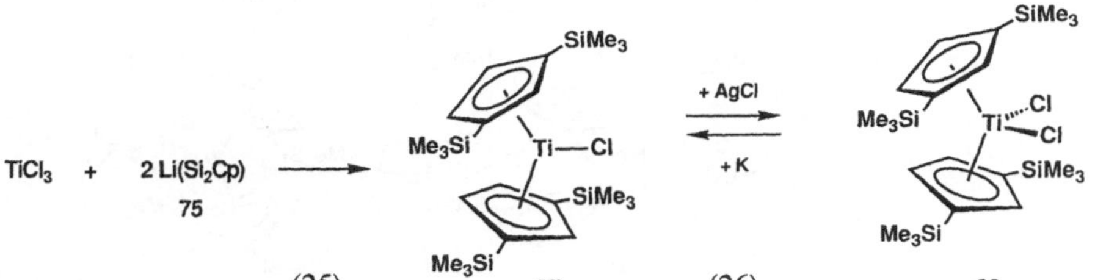

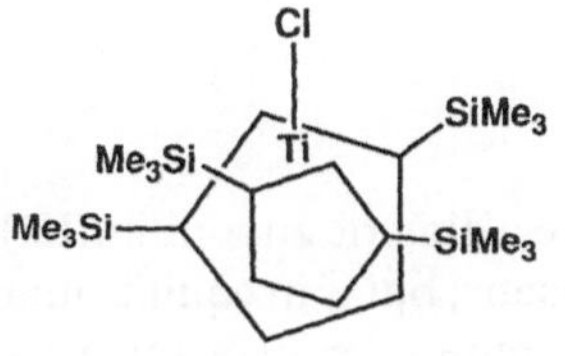

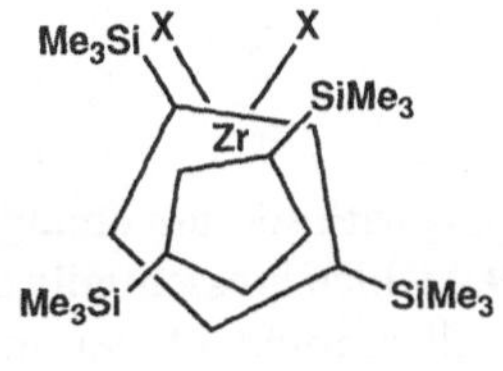

The titanocene dichloride complex **86** was isolated as orange red crystals and found to show by variable-temperature NMR spectroscopy restricted rotation of the ring ligands at low temperatures. The barrier to ring rotation was estimated to be 8.9 kcal mol^{-1} [99]. In the absence of a crystal structure determination one has to assume for **86** that the ground state has C_2 symmetry, since all enantiotopic groups become inequivalent at the low-temperature limit. As shown below, the zirconium dihalide complexes were observed to adopt such a conformation in the crystal lattice. Reduction of **86** with sodium amalgam under CO was found to give the dicarbonyl $Ti(Si_2Cp)_2(CO)_2$ (**89**).

The zirconocene complex $Zr(Si_2Cp)_2Cl_2$ (**90**) is a versatile starting material for a variety of zirconocene complexes with zirconium in the oxidation state IV, III, and II (Scheme 4) and is easily accessible by the reaction of **75** with $ZrCl_4$. The complete series of dihalides were also prepared and except for the dichloride **90**, characterized by X-ray structural analysis [98]. They all show molecular C_2 symmetry with significant longer zirconium-halogen bond and averaged metal-centroid distance. By variable-temperature NMR spectroscopy no sign of restricted rotation was found for $Zr(Si_2Cp)_2Cl_2$ down to $-90\,°C$ [114].

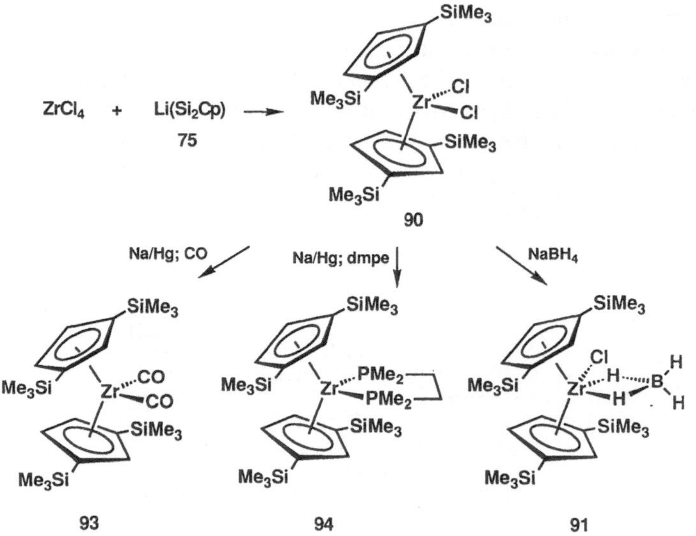

Scheme 4

Reduction of $Zr(Si_2Cp)_2Cl_2$ with sodium amalgam gave a dimeric zirconium(III) complex $[Zr(Si_2Cp)_2Cl]_2$ (**92**) which is essentially diamagnetic, but is in equilibrium even in toluene with a small amount of paramagnetic monomer detectable by EPR spectroscopy [101]. On the other hand, reduction of $Zr(Si_2Cp)_2Cl_2$ with a variety of reductants gave under CO the dicarbonyl $Zr(Si_2Cp)_2(CO)_2$ (**93**) as dark green crystals [102] and reduction using sodium amalgam in the presence of dmpe afforded dark green $Zr(Si_2Cp)_2(dmpe)$ (**94**) which was characterized by X-ray structural analysis [87]. A slow exchange of the bridging and terminal hydrogen atoms of the η^2-bonded BH_4 ligand was observed in $Zr(Si_2Cp)_2Cl(BH_4)$ (**91**) by ^{11}B NMR spectroscopy [87].

Cyclic voltammetry has shown for the series of zirconocene dichloride complexes of the general type $Zr(Si_nCp)_2Cl_2$ that ease of reduction falls off in the sequence $Si_2Cp > SiCp > Cp$ reflecting the electron withdrawing nature of the trimethylsilyl ring-substituents [87]. A similar trend was observed for the niobocene dichloride complexes $Nb(Si_nCp)_2Cl_2$ which serves as a starting material for a number of niobium complex of the formulae $Nb(Si_2Cp)_2(L)Cl$ with $L = CO, RNC,$ O_2 [88].

4.3 [1,2,4-Tris(trimethylsilylcyclopentadienyl)]metal Complexes

4.3.1 Complexes of Iron, Cobalt, and Nickel

Reaction of $Li(Si_3Cp)$ (77) with $FeBr_2$ was reported to give 1,1',2,2',4,4'-hexa-kis(trimethylsilyl)ferrocene $Fe(Si_3Cp)_2$ (95) in low yield [103]. Detailed investigation of the reaction of 77 with ferrous halides FeX_2 (X = Cl, Br, I) led to the discovery that at low temperatures a highly reactive, thermally labile intermediate 96 with only one Si_3Cp ligand per iron center is formed [104]. This species, though not fully characterized, behaves as a functional equivalent of the 14-electron half-sandwich fragment $Fe(Si_3Cp)X$ and can be trapped with two-electron donor L such as CO and $P(OMe)_3$ to give crystalline complexes of the type $Fe(Si_3Cp)L_2X$ (97) [104, 105]. In the presence of $AlCl_3$ and an arene such as toluene it gives rise to the formation of the mixed sandwich complex $[Fe(Si_3Cp)(\eta^6\text{-arene})]^+$ (98) [106]. When 96 is reacted with another cyclopentadienyl transfer reagent such as NaCp or LiCp* mixed ferrocene derivatives $Fe(Si_3Cp)Cp$ (99a) and $Fe(Si_3Cp)Cp*$ (99b) are obtained (Scheme 5) [107].

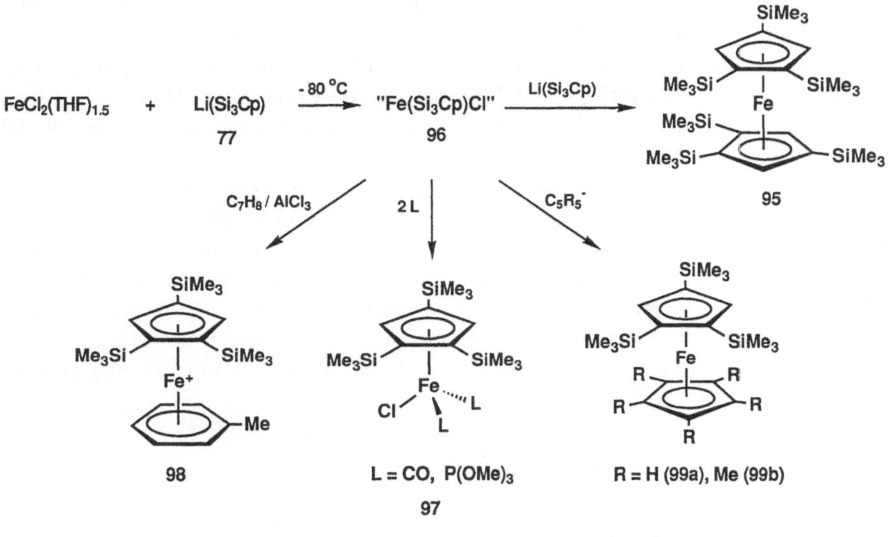

Scheme 5

The low yield of 95 was accounted for by the difficulty of attaching the second Si_3Cp ligand at the iron center due to the extreme bulk of this ligand. However, once formed 95 is isolated as thermally robust, pentane-soluble, red crystals. The sandwich structure was confirmed by an X-ray structral diffraction study [107] showing a rather congested molecule with iron-ring distances of average 2.082 Å and a slight tilt of the two rings by 6.1°. The conformation found in the crystal 95A is clearly dictated by the interannular steric repulsion of the trimethylsilyl groups a pair of which has to be eclipsed so as to enable the other two pairs to be fully staggered.

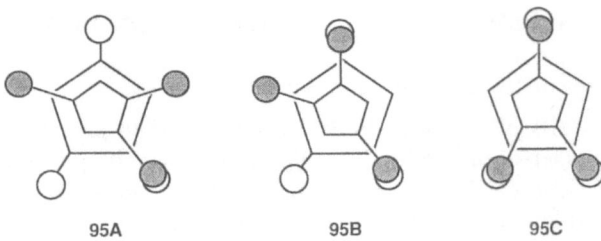

95A   95B   95C

The dynamic property of $Fe(Si_3Cp)_2$ in solution was studied by NMR spectroscopy and revealed a lowering of the molecular symmetry from C_{2h} or C_{2v} to C_2 due to restricted ring rotation at lower temperatures [107]. The associated activation energy of $\Delta G^+ = 11.1$ kcal mol^{-1} was found to be similar to that obtained for the tetrakis(trimethylsilyl)ferrocene 78 and this was explained by assuming an oscillatory motion by 216° rather than a full rotation as postulated in the case of 78. A transition state with all three pairs of trimethylsilyl groups being eclipsed (conformation 95C) appeared to be energetically prohibitive. The fluxional process corresponds to a racemization process of the rotamer 95A and its enantiomer via 95B. A similar rotational barrier of $\Delta G^+ = 9.7$ kcal mol^{-1} was found for $Fe(BuSi_2Cp)_2$ with a slightly modified ligand system [108]. Oxidation of 95 with $AgBF_4$ yielded the blue ferrocenium cation $[Fe(Si_3Cp)_2]BF_4$ that has been characterized by X-ray structural analysis and shown to adopt the conformation 95A [89].

The easily prepared half-sandwich complexes of the type $Fe(Si_3Cp)L_2X$ (97) with $L = P(OMe)_3$ was found to react rapidly with CO to give chiral complexes of the type $Fe(Si_3Cp)L(CO)X$ (100) which according to NMR spectra preferentially adopt a conformation with minimum steric repulsion within the ligand set. Because of the inherent asymmetry it could not be decided whether the ring rotation is hindered [105]. However, a low temperature NMR spectroscopic study of $[Fe(Si_3Cp)L_3]^+$ (101) with $L = P(OMe)_3$ showed no sign of an exchange process being slowed down [106].

SiMe₃   SiMe₃

(MeO)₃P   CO   L   L

Fe   Fe⁺

Me₃Si   I   SiMe₃   Me₃Si   L   SiMe₃

L = P(OMe)₃

100   101

In contrast to the difficulty involved with the formation of ferrocene 95 the cobaltocene analogue $Co(Si_3Cp)_2$ (102) could be straightforwardly synthesized from $CoCl_2$ and $Li(Si_3Cp)$ and isolated as black crystals in good yield. Oxidation of 102 yielded yellow orange cobaltocenium cation $[Co(Si_3Cp)_2]^+$ (103) that as expected showed restricted rotation at lower temperatures. The activation energy for **103** was $\Delta G^{\ne} = 9.9$ kcal mol^{-1} [93], whereas for analogous $[Co(BuSi_2Cp)_2]^+$ a value of $\Delta G^{\ne} = 8.8$ kcal mol^{-1} was determined [108].

M = Co (**102**), Co$^+$ (**103**), Ni (**104**)

105

The formation of the green nickelocene $Ni(Si_3Cp)_2$ derivative **104** from nickel halide and $Li(Si_3Cp)$ proceeded similarly smoothly. Although the intermediate with one Si_3Cp ligand — probably an analogue of the complex $[Ni(\eta^5-C_5Ph_5)X]_2$ — could not be isolated, its existence in the reaction mixture was indicated by the isolation of purple phosphine complexes of the type $Ni(Si_3Cp)LX$ (**105**) (L = PPh$_3$). Variable temperature NMR spectroscopic data for the chloride for instance showed also the restriction of a fluxional process which is caused by the inability of the PPh$_3$ ligand to pass the side of the Si_3Cp ligand with two vicinal trimethylsilyl groups. The rotational barrier to the rotation of the Si_3Cp ligand in **105** was estimated to $\Delta G^{\ne} = 10.5$ kcal mol^{-1} [109].

4.3.2 Complexes of Early Transition Metals

Reaction of $Li(Si_3Cp)$ with titanium(III) chloride resulted in the intermediate formation of $Ti(Si_3Cp)Cl_2$ that was oxidatively converted to the orange trichloride complex $Ti(Si_3Cp)Cl_3$ (**106**), a versatile starting compound for titanium half-sandwich complexes bearing a Si_3Cp ligand (Scheme 6) [111]. The analogue of **106** with the $BuSi_2Cp$ ligand $Ti(BuSi_2Cp)Cl_3$ has also been isolted as orange red crystals [110]. All attempts to attach two Si_3Cp ligands at titanium have failed so far. This failure may be due to the extreme strain two Si_3Cp ligands cause at a transition metal center. The solid state structure of the 1,2,4-tris(trimethyl-silyl)titanocene dichloride $Ti(Si_3Cp)CpCl_2$ (**107**) was determined and revealed a slight distortion of the Si_3Cp ligand toward an η^3-coordination [112]. The conformation of this complex illustrates the interplay of specific steric repulsion within a titanocene framework. Thus, the bending angle of 129.7° indicate an interference of one SiMe$_3$ group with the chlorine atoms and is probably responsible for the distortion toward an η^3-bonding. As has been noted in other highly substituted titanocene complexes such as **20**, the side opposite to the open cleft seems to be the least favored position for the ring substituents.

The hydrolysis of **106** was studied in some detail and led to the isolation and structural characterization of a rare example of a titanoxane derivative with a planar $Ti_2(\mu-O)_2$ core **108** [111]. The formation of complex **108** can be contrasted with the hydrolysis reaction of the trihalides in the Cp and Cp* series giving rise

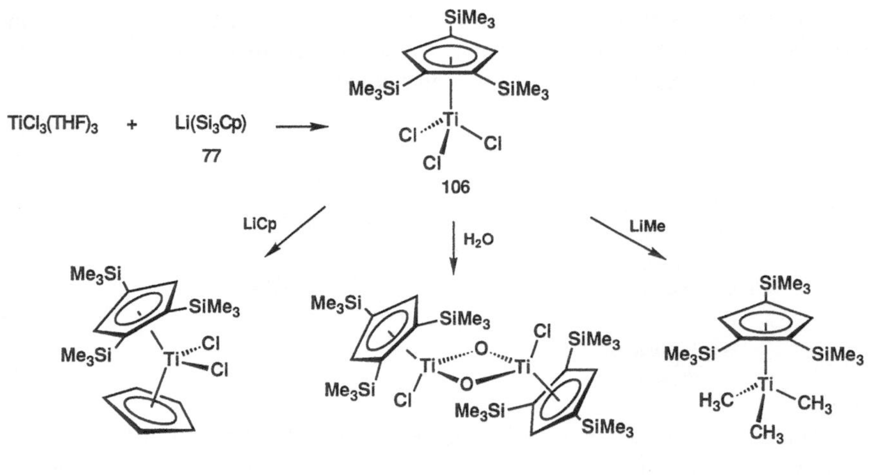

Scheme 6

to tetra- and trinuclear oxo complexes [113]. Apparently the aggregation/condensation reaction associated with hydrolysis of titanium halide complexes is strongly dependent on the nature of the ring-substituent. Another example for a sterically induced stability constitutes the trimethyl complex $Ti(Si_3Cp)Me_3$ (109) that can be straightforwardly synthesized from the trichloride 106 with LiMe and that as compared to the parent complex $TiCpMe_3$ is remarkably thermally robust. The Cp^* analogue $TiCp^*Me_3$ exhibits a similar thermal stability [110].

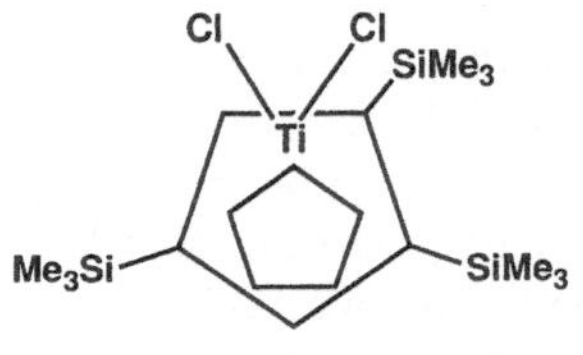

107

A highly congested zirconium complex bearing a Si_3Cp ligand, $Zr(Si_3Cp)$ $(NMe_2)_3$ (110), was synthesized by the amine elimination method using $Zr(NMe_2)_4$ and 76 according to Eq. 27. The ^{13}C NMR spectrum at $-53\ ^\circ C$ is characterized by three different NMe_2 groups indicating the presence of a molecular plane of symmetry. This leads to a conformation in which one NMe_2 ligand is different from the other two. According to the structure in the crystal one amido ligand is positioned directly under the side of the Si_3Cp ligand with two vicinal trimethylsilyl groups. Its methyl groups are not skewed like the others, consequently becoming equivalent, whereas the two other NMe_2 ligands are canted significantly in order to accomodate one methyl group close to the unsubstitued ring carbon.

$$\text{Zr(NMe}_2)_4 \; + \quad \underset{\textbf{76}}{\text{Me}_3\text{Si} \diagdown} \begin{array}{c} \diagup \text{SiMe}_3 \\ \diagdown \text{SiMe}_3 \end{array} \quad \longrightarrow \quad \underset{\textbf{110}}{\boxed{\text{complex 110}}} \qquad (27)$$

The metallocene dichlorides bearing two Si_3Cp ligands, $M(Si_3Cp)_2Cl_2$ (**111**) (M = Zr, Hf), were synthesized starting from $Li(Si_3Cp)$ and the metal tetrahalide (Eq. 28) [114]. The ^{1}H NMR spectra at lower temperatures show restricted rotation with activation barriers of 11.2 and 11.0 kcal mol^{-1} for the zirconocene **111a** and hafnocene derivative **111b**, respectively. It was noted that the analogous complexes derived from the Si_2Cp ligand **90** has a rotational barrier ≤ 9.2 kcal mol^{-1}, indicating a substantial increase in rigidity upon placing six trimethylsilyl groups into the heavier metallocene unit. The observed data are consistent with a full rotation occuring at higher temperatures, in contrast to the ferrocene complex **95**, where only partial rotation was postulated.

$$\text{MCl}_4 \; + \quad \underset{\textbf{77}}{2 \, \text{Li(Si}_3\text{Cp)}} \quad \longrightarrow \quad \underset{\substack{\textbf{111} \\ \text{M = Zr (a) , Hf (b)}}}{\boxed{\text{complex 111}}} \qquad (28)$$

5 Hindered Rotation of the Cyclopentadienyl Ligands

The barrier to rotation of the cyclopentadienyl ligand about the metal-ligand axis was historically first considered using substituted ferrocene derivatives [115] and, by the absence of rotational isomers, concluded to be low. Subsequently a variety of physical techniques were employed to quantitatively determine the activation energy for this process. For the unsubstituted cyclopentadienyl ring the barrier in both metallocene and half-sandwich complexes are invariably low and seems to be dominated by intermolecular interactions rather than internal electronic or steric constraints [116]. Only very recently was reported the first example of a unsubstituted cyclopentadienyl ring showing restricted rotation that can be studied by conventional variable-temperature NMR spectroscopy in solution [117]. In $[OsCp(PPh_3)_2(PhCH=CH_2)]^+$ (**112**) this observation was accounted for by an

129

Table 1. Energy barriers to ring rotation in unsubstituted and substituted cyclopentadienyl transition metal complexes.

Ring	Complex	Method	E_a [kcal mol^{-1}]	Ref.
C_5H_5	$Fe(\eta^5\text{-}C_5H_5)_2$	A1	1.8–2.3	116b–d
C_5H_5	$M(\eta^5\text{-}C_5H_5)_2$ $M = Co, Co^+, Ni$	A1	1.8	116a
C_5H_5	$Ru(\eta^5\text{-}C_5H_5)_2$	A1	2.3	116b
C_5H_5	$[Os(\eta^5\text{-}C_5H_5)\text{-}(PPh_3)_2L]^+$ (L = PhCH=CH$_2$) (**112**)	C	8.2	117
C_5H_5	$Ti(\eta^5\text{-}C_5H_5)_2Cl_2$	A1	0.5	116b
C_5H_5	$Ti(\eta^5\text{-}C_5H_5)Cl_3$	A1	2.3	116e
C_5H_5	$Rh(\eta^5\text{-}C_5H_5)(COD)$	A2	1.9	116i
C_5H_5	$Pt(\eta^5\text{-}C_5H_5)(CH_3)_3$	B	4.9	116j
BuCp	$Fe(BuCp)_2$	A1	6.8	116d
Bu_2Cp	$Fe(Bu_2Cp)_2$ (**6**)	C	13.1	31
Bu_2Cp	$[Co(Bu_2Cp)_2]^+$ (**18**)	C	12.7	35
Pr_4Cp	$Fe(Pr_4Cp)_2$ (**26**)	C	13.6	43b
Si_2Cp	$Fe(Si_2Cp)_2$ (**78**)	C	11.0	92
Si_2Cp	$[Co(Si_2Cp)_2]^+$ (**81**)	C	10.7	93
Si_2Cp	$Ti(Si_2Cp)_2Cl_2$ (**86**)	C	8.9	99
Si_3Cp	$Fe(Si_3Cp)_2$ (**95**)	C	11.0	107
Si_3Cp	$[Co(Si_3Cp)_2]^+$ (**103**)	C	9.9	93
Si_3Cp	$Zr(Si_3Cp)_2Cl_2$ (**111a**)	C	11.0	114
Si_3Cp	$Hf(Si_3Cp)_2Cl_2$ (**111b**)	C	11.3	114
Si_3Cp	$Ni(Si_3Cp)(PPh_3)Cl$ (**105**)	C	10.5	109
$BuSi_2Cp$	$Fe(BuSi_2Cp)_2$	C	9.7	108
$BuSi_2Cp$	$[Co(BuSi_2Cp)_2]^+$	C	8.8	108
C_5HPh_4	$Mo(\eta^5\text{-}C_5HPh_4)(CO)_2L_2$ $L_2 = O(CO)_2C_2(PPh_3)_2$ (**44**)	D	9.9	65

A: Relaxation time measurement in the solid (A1); in solution (A2). B: Mechanical spectroscopy. C: Variable-temperature NMR spectroscopy (coalescence temperature measurement). D: Variable-temperature EPR spectroscopy

extreme steric congestion around the metal. It is noteworthy that in Cp* complexes also no notable increase in the rotational barrier can be found [118].

From the data presented in Table 1 it becomes clear that attaching sterically demanding ring-substituents significantly enhances the conformational rigidity and results in rotational barriers measureable in solution by variable-temperature NMR spectroscopy. The increase of activation energy for the ring rotation in highly substituted metallocenes by about an order of magnitude should have some consequence for the design of "stereorigid" transition metal complexes.

Although any generalization seems to be premature at the present time, it is interesting to note that tetra- and pentaphenylcyclopentadienyl ligands have not allowed but for one case the observation of hindered cyclopentadienyl-ring rotation. This is accounted for in a plausible manner by considering the propeller-like rotation of "flat" phenyl rings. There is substantial material concerning the restricted rotation of substituted arene ring in bis- [119] and mono-arene transition metal complexes [51 b, 120] that show close similarity to cyclopentadienyl metal complexes under consideration here. In the case of half-sandwich complexes of the type $Ru(\eta^6\text{-arene})(SiCl_3)(CO)_2$ a transition state in which ring tilting was impossible due to the substitution pattern was made responsible for the observability of restricted rotation [120]. Extrapolation of this hypothesis to complexes with cyclopentadienyl ligands infers that the presence of at least two bulky substituents in the 1,3-position is a minimal requirement for a restricted rotation to be observed in parallel metallocene and half-sandwich complexes.

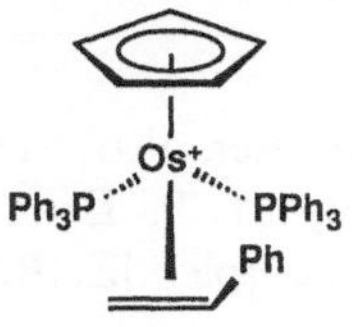

112 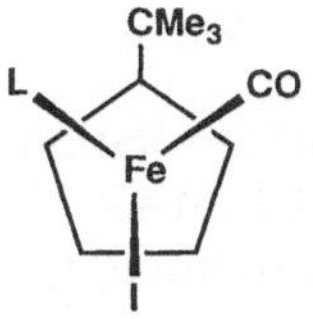   113

Conformational features appear to be even more delicately balanced in bent metallocenes where the rotational profile is complicated by the fact that in addition to interannular repulsive interaction of the ring-substituents close contacts to the additional ligands on the open side becomes more significant. Consequently, no systematics for the adoption of a certain conformation in multiply ring-substituted bent metallocene can be deduced so far. The situation, however, becomes clearer when only mono-substituted cyclopentadienyl ligands are considered. While for half-sandwich complexes of the type Fe(BuCp)CO(L)I (113) conformational preferences could be extracted from detailed NMR spectroscopic studies in conjunction with molecular modeling studies [121], zirconocene derivatives with two BuCp ligands show evidence for restricted rotation depending on the nature of the additional ligands coordinated at the zirconium center. Erker has thoroughly performed an investigation into this effect and determined the rotational barrier as a function of both the additional ligands and the ring-substituent for a series of complexes of the type $Zr(BuCp)_2L$ such as 114 which are summarized in Table 2. Related complexes including the dichloride $Zr(BuCp)_2Cl_2$ that does not allow

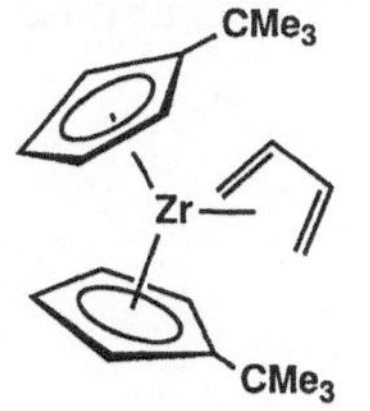

114

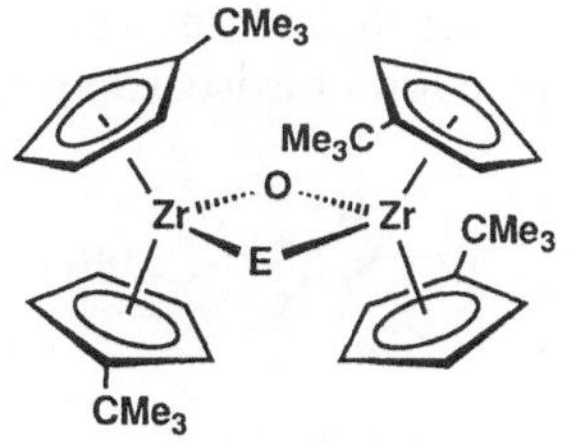

E = Se, Te
115

131

Table 2. Activation energy for ring rotation in 1,1′-disubstitued zirconocene complexes

R	L	$\Delta G^{\neq}$, kcal mol^{-1} (Tc, °C)	Ref.
CMe$_3$	s-cis-C$_4$H$_6$	9.8 (−86)	121a
CMe$_3$	s-cis-C$_4$H$_6$	9.2 (−77)	121b
CMe$_3$	s-cis-2,3-C$_4$H$_4$Me$_2$	8.4 (−97)	121b
CMe$_3$	s-trans-2,3-C$_4$H$_4$Me$_2$	8.6 (−100)	121b
H, CMe$_3$	s-cis-C$_4$H$_6$	9.4 (−75)	121b
CMe$_2$(n-C$_4$H$_9$)	s-cis-C$_4$H$_6$	10.1 (−64)	121c
CMe$_2$(n-C$_{10}$H$_{21}$)	s-cis-C$_4$H$_6$	10.1 (−63)	121c
CMe(CH$_2$)$_5$	s-cis-C$_4$H$_6$	10.2 (−62)	121c

the observation of restricted rotation in solution have been also examined in the solid state using ^{13}C CP/MAS NMR spectroscopy [122]. Examples for an intramolecular hindered rotation within a dinuclear complex [Zr(BuCp)$_2$]$_2$(μ-O) (μ-E) (**115**) (E = Se, Te) were also reported [123].

6 Chiral and Optically Active Cyclopentadienyl Ligands

Because of the great potential application organotransition metal complexes have in both stoichiometric and catalytic asymmetric synthesis [124, 125], the development of methods for rationally positioning transition metal centers within defined chiral ligand spheres is an objective of increasing importance. In contrast to the role alkoxy and phosphine ligands [126] play in the design of efficiently working homogeneous asymmetric catalysts, cyclopentadienyl ligands have not gained much popularity as ancillary ligands for the control of stereoselectivity during a metal-induced transformation. This is somewhat surprising if one considers that stereochemical aspects, especially chirality associated with cyclopentadienyl complexes, have been studied quite extensively in the past [127]. The fact that two different ring-substituents on a cyclopentadienyl ligand induce a molecular asymmetry when attached to a metal center was recognized soon after the discovery of ferrocene and the first resolution of a chiral ferrocene was achieved as early as in 1959 [128]. The concept of planar chirality in cyclopentadienyl-metal complexes was coined metallocene chirality and has been the subject of comprehensive reviews [129]. Chiral ligand such as **116** based on this concept has recently been developed and successfully applied in the enantioselective gold-catalyzed aldol reaction of α-isocyanoacetates [130].

116

In tetrahedral complexes with a stereogenic metal center the cyclopentadienyl ligand may constitute one of the four different ligands. In recent years, a number of half-sandwich fragments of the general type $MCpL_2$ or $MCpX_2$ (e.g. $ML_2 = Fe(CO)(PPh_3)$ [131], $Re(NO)(PPh_3)$ [132], $[Mo(CO)(NO)]^+$ [133]; $MX_2 = Ti(OR*)_2$ [134]) have gained considerable importance as transition metal-centered chiral auxiliary for various highly efficient stereoselective reactions. In these systems the cyclopentadienyl ligand has some beneficiary effects on the configurational stability of the coordination geometry around the chiral transition metal center, but does not apparently play an "active" role in the metal-centered reaction. Some developments in the design and complexation of chiral and, in particular, optically active substituted cyclopentadienyl ligands are summarized below.

6.1 Cyclopentadienyl Ligand with One Chiral Substituent

There are a number of studies aimed at chirally modifying cyclopentadienyl ligands by introducing an optically active substituent, conceptually the most simple method to arrive at optically active cyclopentadienyl-metal complexes [135]. Mono-substituted cyclopentadienyl ligands of the type C_5H_4R* have been described, where $R*$ is a resolvable chiral group such as menthyl (a), neomenthyl (b), phenylmenthyl (c), or 1-phenylethyl (d). These ligand systems have successfully been employed to ensure diastereoselectivity in reactions involving chiral metal centers [136]. However, it remains unclear to what extent, either by electronic or steric factors, the peripheral substituents influence the stereochemical course of a metal-centered transformation. In general, it is agreed on that a chiral and sterically somewhat encumbered, less conformationally flexible ("stereorigid") reaction site is required to induce a high degree of stereoselection [137].

A catalytic hydrogenation using titanocene complexes of the type $Ti(\eta^5\text{-}C_5H_4R*)_2Cl_2$ (117) and $Ti(\eta^5\text{-}C_5H_4R*)CpCl_2$ (119) carrying optically active menthyl- and neomenthyl-cyclopentadienyl ligands (Eq. 29) showed some enantio-selectivity when 2-phenylbutene was used as a prochiral substrate molecule [138]. Likewise, carboxylation of allyl titanium(III) complexes $Ti(\eta^5\text{-}C_5H_4R*)\text{-}(\eta^3\text{-}H_2CCRCHCH_3)$ 118a gave after hydrolysis chiral 2-butenoic acid $H_2C=CRCH(CH_3)CO_2H$ with some enantioselectivity [139]. The moderate selectivity during these reactions is partly accounted for by the molecular structure of $Ti(\eta^5\text{-}C_5H_4R*)CpCl_2$ (119a) that shows a fairly unexceptional structure of a titanocene dichloride derivative with laterally positioned $\eta^5\text{-}C_5H_4R*$ ligand [140]. The ligand set is probably incapable of providing a sufficiently rigid asymmetric environment for diastereofacial differentiation, although by variable-temperature NMR spectroscopy a fluxional behavior was observed. This interesting feature, however, was not conclusively examined nor interpreted.

Hydrido complexes of similarly modified zirconocene derivatives $[Zr(\eta^5\text{-}C_5H_4R*)CpH_2]_2$ ($R* = CH_2CHMeEt$, $CH_2CHPhEt$) 120 gave virtually no enantioselection when used as hydrogenation catalyst for prochiral olefins, although the activity was quite high [141].

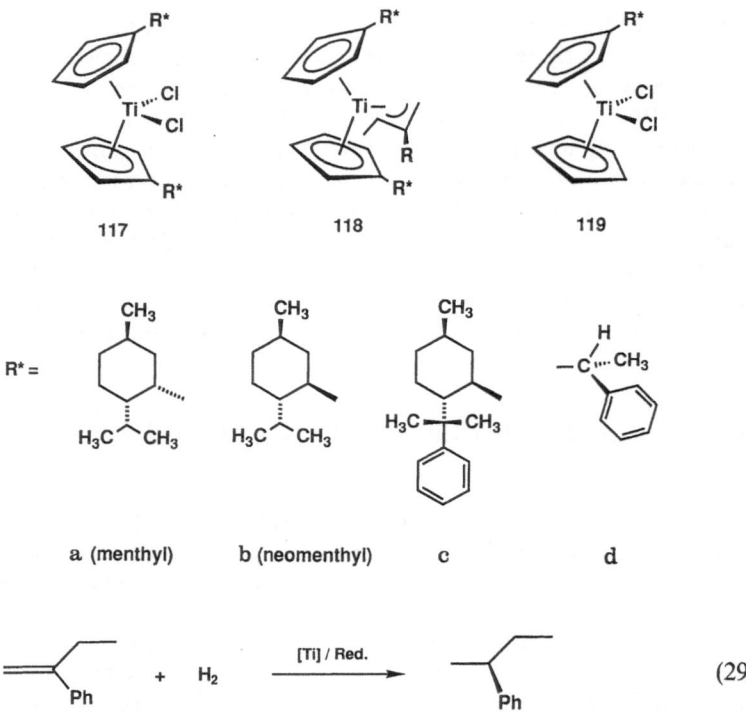

$$R^* =$$

a (menthyl)   b (neomenthyl)   c   d

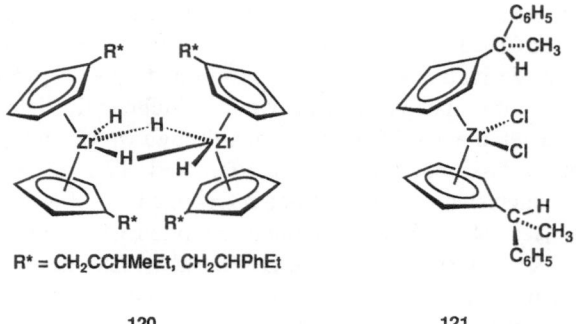

(29)

Zirconocene dichloride **121** derived from (1-phenylethyl)cyclopentadienyl ligand is formed as a mixture of diastereomers from which the racemic form can be isolated by fractional crystallization. This complex was studied by X-ray diffraction methods and revealed a virtually chiral C_2-symmetrical conformation in which the chiral ring-substituents are arranged in a synclinal position relative to the five-membered ring. It was proposed that this conformation is preserved in solution. Using **121** as catalyst the influence of double stereodifferentiation during isospecific polymerization of propylene (Eq. 32) was demonstrated for the first time [142].

R* = CH₂CCHMeEt, CH₂CHPhEt

120   121

6.2 Optically Active Ring-Annulated Cyclopentadienyl Ligands

6.2.1 C_1-Symmetrical Ligands

When an asymmetric group is attached on the periphery of the cyclopentadienyl ligand, the two sides of the five-membered ring becomes diastereotopic. Thus, coordination of an achiral transition metal fragment should give rise to two isomers. While for cyclopentadienyl ligands with one chiral ring-substituent such as 1-phenylethyl the two diastereomers obtained are rotamers and therefore interconvertible by rotation about the σ-bond, two pairs of diastereomers are inevitably formed when the chiral ring substituent is part of an annulated framework attached to the five-membered ring (which in principle is equivalent to the case where the cyclopentadienyl ligand bears two different ring substituents one of which is chiral). For an achiral annulated cyclopentadienyl ring the coordination of a transition metal fragment may occur diastereoselectively. Although for this so-called π-facial selectivity electronic factors seem to be dominant, steric interaction between the ligands can not be neglected [143]. This problem was analyzed to some depth using isodicyclopentadienyl ligands **122** and is briefly mentioned here, since these tricyclic ligands represent a model for optically active polycycle-annulated cyclopentadienyl systems.

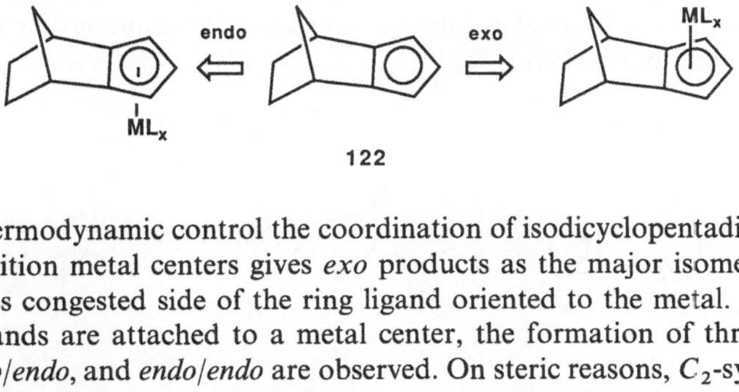

122

Under thermodynamic control the coordination of isodicyclopentadienyl anion **122** at transition metal centers gives *exo* products as the major isomer with the sterically less congested side of the ring ligand oriented to the metal. When two of these ligands are attached to a metal center, the formation of three isomers *exo/exo*, *exo/endo*, and *endo/endo* are observed. On steric reasons, C_2-symmetrical *exo/exo* isomer is favored. Structurally characterized examples confirms this finding. Thus, the ferrocene derivative of the dehydro ligand system **123** exhibits a conformation in which the five-membered rings are fully eclipsed and the ring-substituents arranged "gauche" [144].

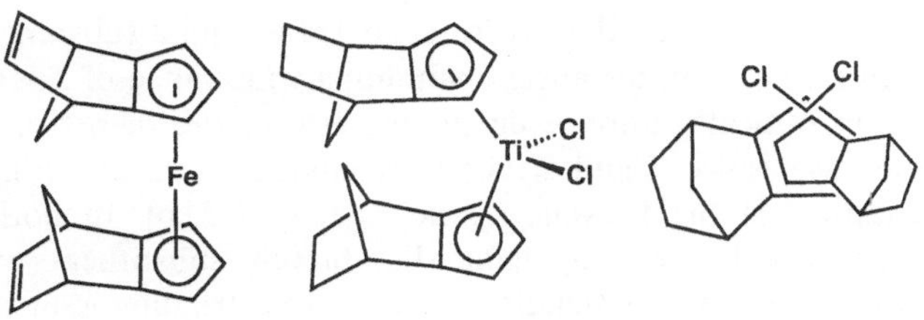

123 **124**

M = Ti (a), Zr (b)

The titanocene dichloride derivative **124a** was synthesized using TiCl$_3$ and the isodicyclopentadienyllithium and its structure determined [145]. The homologous zirconocene derivative **124b** shows the same structural feature in the solid state. In the case of titanocene complexes *endo* coordination could be also observed under kinetic control, when TiCpCl$_3$ or TiCp'Cl$_3$ was reacted with isodicyclopentadienyllithium at low temperatures, whereas at room temperature the same reaction gave products with *exo* coordination at the titanium center [146].

A cyclopentadienyl ligand with an optically active group annulated to the five-membered group were first reported in 1985 [147]. Apart from the camphor-derived system **125** related optically active cyclopentadienyl ligands derived from pinene **126** [148] and verbenone **127** [151] have been synthesized. The coordination of these ligands to transition metal centers usually entails diastereomer formation with comparable selectivity as found for the isodicyclopentadienyl system **122** [148]. Thus, a mixture of diastereomers was formed when the Co(CO)$_2$ fragment was attached to the camphor-derived system **125** [147, 149]. The major isomer of the zirconocene dichloride complex of the camphor-derived ligand **128b** show a conformation analogous to that of the bis(isodicyclopentadienyl)zirconium complex **124b**, with the peripheral framework oriented laterally to avoid interligand repulsion [150a]. Similarly, pinene-derived ligand system **126** led to the titanocene dichloride derivative **129** of C_2 symmetry with a very similar conformation [150b]. The dimeric iron complex **130** with the ligand **126** is remarkable in that *cis*-conformation is observed in the solid state, but no study of its dynamic behavior in solution was undertaken [148].

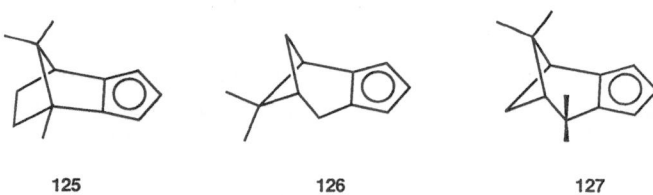

125	126	127

With the more encumbered verbenone-derived ligand system **127** the titanocene dichloride complex **131** exhibits a rather constrained ligand sphere around the titanium center [151a]. The bicyclic groups are oriented in such a manner that they are placed above and below the TiCl$_2$ fragment. This leads to the decrease of the bending angle between the two rings to 131.9° and a substantial difference in the titanium ring carbon distances spanning a wide range of 2.31(1) to 2.63(1) Å [151a]. Using optically impure ligand precursors the *meso* form **132** of C_s symmetry was also isolated and structurally characterized revealing a different mutual orientation of the tricyclic ligand systems [151b]. In both cases it is important to see that the sterically demanding bicyclic substituents avoid the side opposing the open side of the metallocene unit. An intriguing issue would be the possible observability of enantiomer self-recognition if racemic ligand is utilized for complexation.

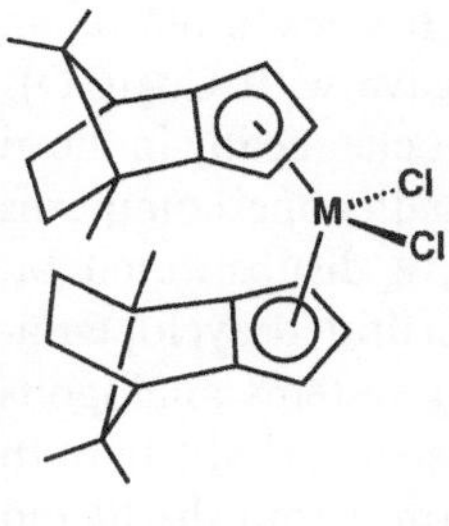

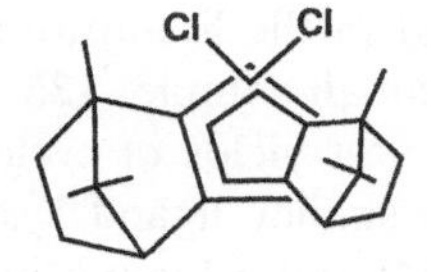

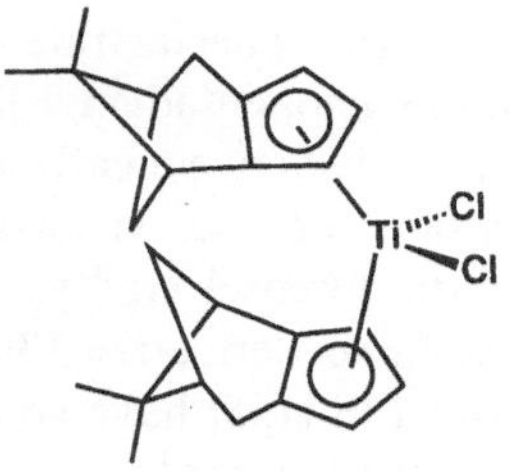

128
M = Ti (a), Zr (b)

129

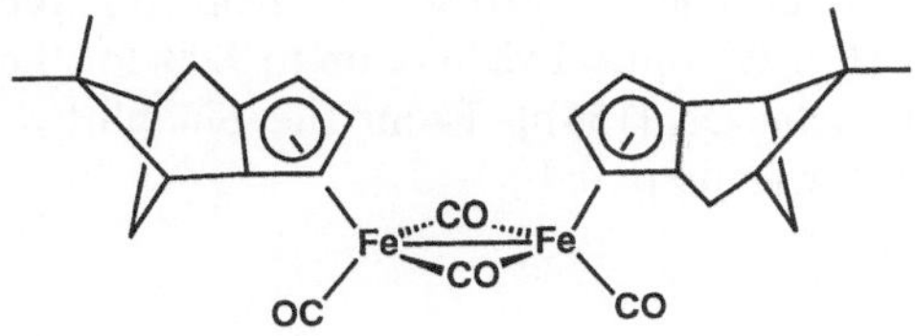

130

A certain π-facial selectivity was achieved when MCpCl$_2$ (M = Ti, Zr) fragments were coordinated to the optically active fused cyclopentadienyl ligands. For instance, reaction of ZrCpCl$_3$ with the lithium derivative of **126** at −78 °C gave predominantly **133** which was characterized by X-ray structural analysis [152].

The titanocene dichloride complexes derived from the camphor- and pinene-annulated ligands **126** and **127** were tested as enantioselective hydrogenation catalyst and using 2-phenylbutene as substrate 2-phenylbutane was obtained with *ee* up to 34% [148, 149].

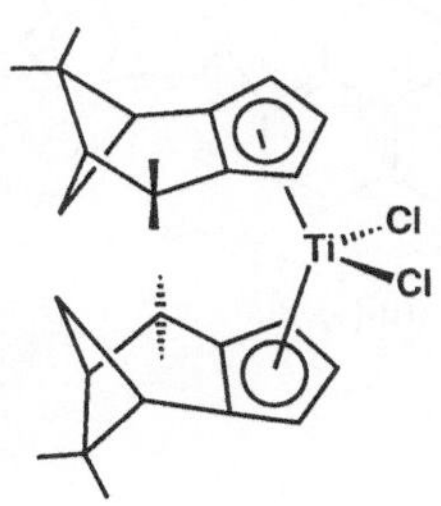

131

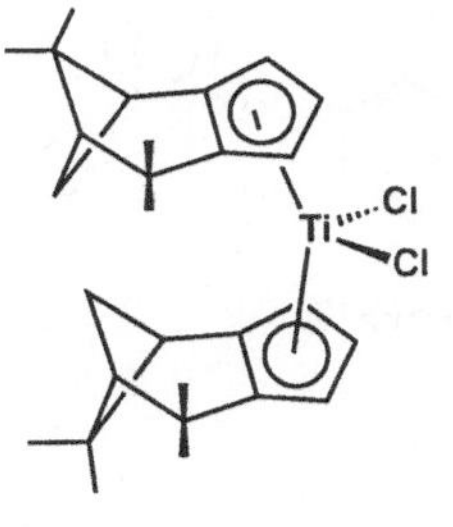

132

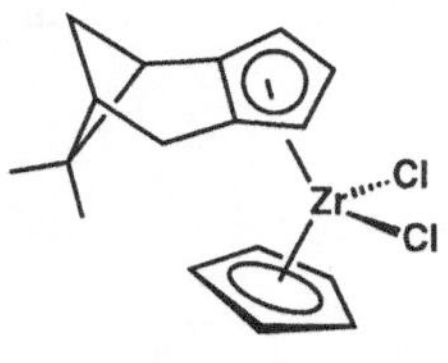

133

6.2.2 C_2-Symmetrical Ligands

The problem of π-facial differentiation, i.e. diastereomer formation, encountered in the metal complexation of the above mentioned annulated cyclopentadienyl ligands is avoided when C_2-symmetrical ligands [153] are utilized. Since in such ligands both sides of the five-membered rings are *homotopic*, only one isomer is expected to form. The tartrate-derived ligand **134** was the first example of an axially chiral cyclopentadienyl ligand in the literature and gave with $Co_2(CO)_8$ only one optically active dicarbonyl cobalt complex **135**. Photochemically induced cyclization of α,ω-diynes yielded modest yields of cyclopentadienone complexes with some diastereoselectivity. Two similar ligand systems, a disubstituted bicyclohexane-fused derivative **136** [154] and a binaphthyl-substituted cyclopentadienyl ligand **138** [155] have so far been synthesized. Ligand systems analogous to **136** with substituents other than phenyl groups are also accessible [154b]. In both cases either enantiomer can be obtained in optically pure form. From the ligand system **136** cobalt carbonyl, titanium, and zirconium complexes were prepared. The asymmetric steric encumbrance around the metal was demonstrated by an X-ray structural investigation of the titanocene derivative **138**. Applying the catalytic test reaction mentioned above (Eq. 29), optical yield of up to 92% for the hydrogenation of 2-phenylbutene was achieved [154b]. From the binaphthyl-derived ligand **138** titanium complex **139** was prepared.

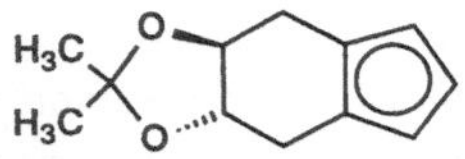

134

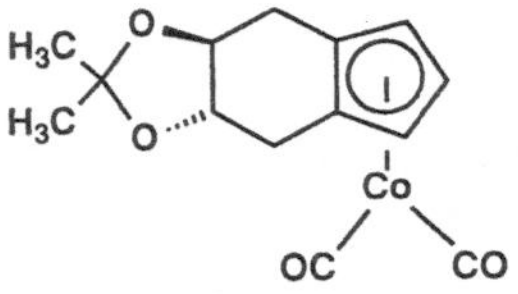

135

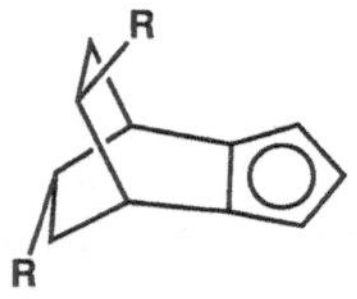

R = Ph, ^{i}Pr

136

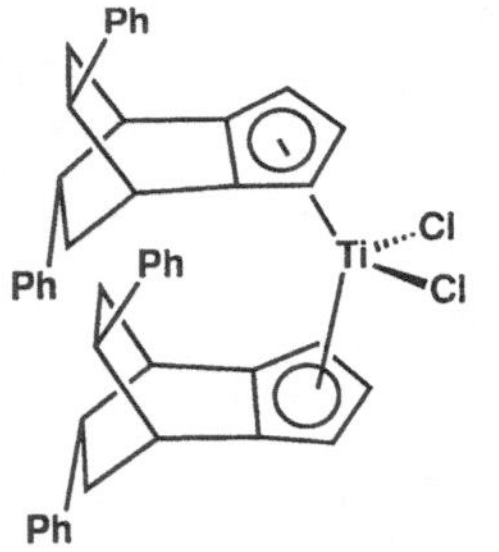

137

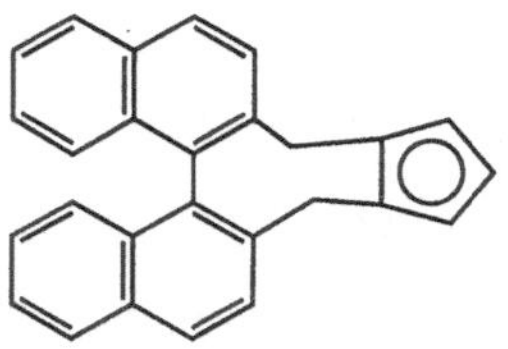

(S)-(-)-138

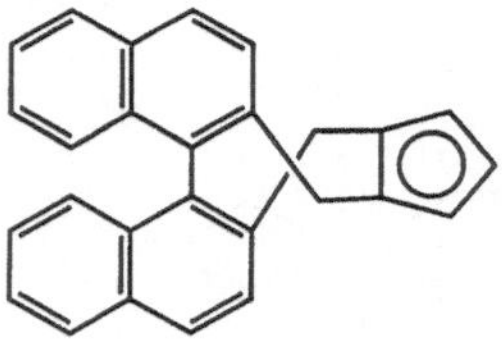

(R)-(+)-138

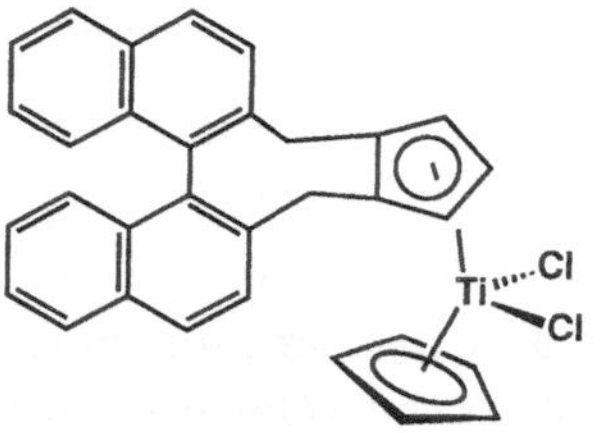

139

An optically active cyclopentadienyl ligand with two annulated rings was prepared starting with either (+)- or (−)-camphor by a sequence similar to one synthesis of the pentamethylcyclopentadiene [156]. After converting the cyclopentadiene **140** to the lithium derivative the ligand was coordinated to the trichloro metal fragment of zirconium and hafnium using the corresponding metal tetrahalide (Eq. 30). The zirconium complex **141** was employed as an efficient chiral Lewis acid in the catalytic reaction of 1-naphthol with pyruvates to give 2-(2-hydroxy-naphthyl)lactate with up to 89% *ee* under optimized conditions (Eq. 31) [156b].

$$\text{140} \quad \xrightarrow[\text{2. MCl}_4]{\text{1. Li}^n\text{Bu}} \quad \text{M = Zr (141), Hf} \tag{30}$$

$$\tag{31}$$

6.3 Linked Cyclopentadienyl Ligands

A report on the C_2-symmetrical *ansa*-zirconocene complex *rac*-dichloro{ethyl-enebis(η^5-tetrahydroindenyl)}zirconium **142** as an highly isospecific catalyst for the homogeneous polymerization of α-olefin (Eq. 32) [157] led to a sudden increase of interest in designing new substituted cyclopentadienyl ligands suitable for "stereorigid" complexation of catalytically active transition metal centers. It was soon recognized that apart from parameters such as the chain length [158] peripheral substituents on the cyclopentadienyl ligand systems play a key role in controlling both catalytic activity and stereoselectivity of the C−C bond formation, offering a good possibility for studying structure-reactivity relationships on a molecular level. The underlying principle for the synthesis and properties of *ansa*-metallocenes have been ingeniously developed by Brintzinger, who in 1979 had succeeded in resolving a chiral *ansa*-titanocene (*S*)-binaphtholate complex **143** [159]. Subsequently a variety of related linked ligand systems have been prepared by Brintzinger and their complexation behavior investigated [160]. Within the scope of the present review such linked cyclopentadienyl and indenyl ligands cannot be considered to the full extent. However, it should be mentioned explicitly that bulky substituents such as *tert*-butyl groups are important for controlling the stereoregularity of polypropylene. For example, the dimethylsilylene-bridged ligand $(\eta^5\text{-}C_5H_2\text{-}2\text{-}Me\text{-}4\text{-}CMe_3)_2SiMe_2$ preferably forms the racemic over the *meso* zirconocene dichloride complex **144** [160] and exhibits comparatively high stereoselectivity when used for isospecific propylene polymerization under standard conditions [161, 162]. A considerable number of similar chelating ligand systems and *ansa*-metallocene dichloride complexes derived therefrom have been reported [163] and a clear picture of the determining influences for creation of a stereorigid coordination site is probably emerging soon.

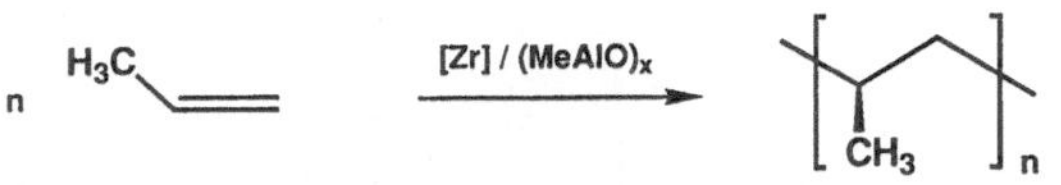

$$(32)$$

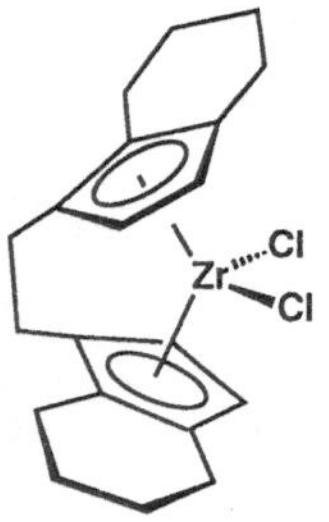

142 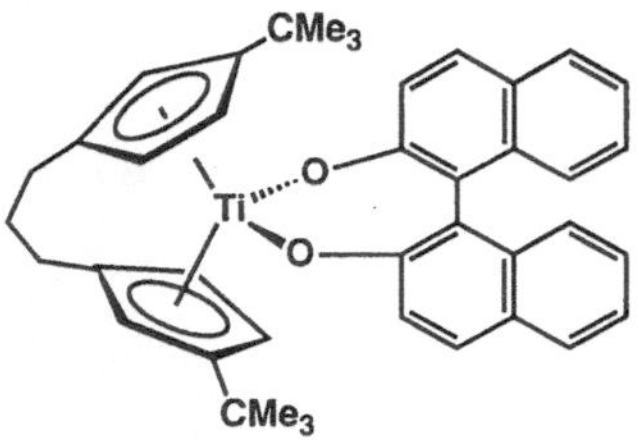   143 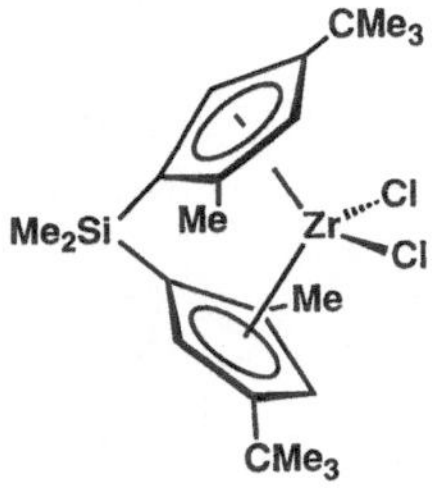   144

7 Conclusion

From the still relatively scattered data on transition metal complexes with sterically demanding cyclopentadienyl ligands one can already recognize a distinct complexation behavior as well as novel types of conformational effects not encountered for the Cp and Cp* analogues. While the complexation of the cyclopentadienyl ligand at a specific transition metal center is still a major synthetic obstacle, a variety of ring-substituents with different steric and electronic properties are available and more complete series of complexes should be accessible in the near future. The precise knowledge of ligand-ligand interactions — both in the solid state by X-ray structure analysis and in solution by NMR spectroscopy — are supposedly of great importance as one is planning rational design of ancillary ligands for transition metal complexes for stereoselective reactions. With the availability of molecular modeling programs conformational analysis of transition metal complexes should become also routine providing a useful tool even for a synthetic chemist.

Addendum

After the completion of this manuscript a paper concerning conformational analyses of 1,1',3,3'-tetra-*tert*-alkylmetallocene of iron and ruthenium including **6** based on thorough NMR spectroscopic measurements (line-shape analysis) has appeared in which the nature of the transition states has conclusively been discussed in detail [164].

Acknowledgement. Our own work on sterically demanding cyclopentadienyl ligands have generously been supported by the Volkswagen-Stiftung, Bund der Freunde der TU München, and the Fonds der Chemischen Industrie. I thank Prof. W. A. Herrmann for his continued interest and encouragement, Prof. M. F. Lappert, Brighton, for kindly providing me with unpublished data on complex **87**, and Prof. N. J. Coville for sending me a preprint of ref. 19.

References

1. a) Kealy TJ, Pauson, PL (1951) Nature 168: 1039; b) Miller SA, Tebboth JA, Tremaine JF (1952) J Chem Soc 632
2. a) Fischer EO, Pfab W (1952) Z Naturforsch 7B: 377; Ruch E, Fischer EO (1952) ibid 7B: 632; b) Wilkinson G, Rosenblum M, Whiting MC, Woodward RB (1952) J Am Chem Soc 74: 2125; Woodward RB, Rosenblum M, Whiting MC (1952) ibid 74: 3458
3. Wilkinson G, Stone FGA, Abel EW (eds) (1982) Comprehensive organometallic chemistry, Pergamon, Oxford
4. Jutzi P (1986) Adv Organomet Chem 26: 217; (1990) J Organomet Chem 400: 1
5. Marks TJ, Fragala IL (eds) (1985) Fundamental and technological aspects of the f-elements; Reidel, Dordrecht, Organometallics of the f-elements, Marks TJ, Fischer RD (eds) (1979) Reidel, Dordrecht
6. Haaland A (1979) Acc Chem Res 12: 415
7. Green MLH, Green JC (In preparation) Systematic chemistry of covalent compounds of the d-block transition metals; Seddon EA, Seddon KR (1984) The chemistry of ruthenium, Elsevier, Amsterdam
8. Lauher JW, Hoffmann R (1976) J Am Chem Soc 98: 1729
9. See for example: Jacobsen DB, Freiser BS (1985) J Am Chem Soc 107: 7399
10. See for example: Döppert K (1990) Naturwissenschaften 77: 19
11. See for example: Okuda J, Herrmann WA, (1987) J Mol Cat 41: 109; Herrmann WA (1988) Comment Inorg Chem 7: 73
12. Ellis JE (1990) Adv Organomet Chem 31: 1
13. O'Connor JM, Casey CP (1987) Chem Rev 87: 307
14. King RB, Bisnette MB (1967) J Organomet Chem 8: 287
15. For examples see: a) Maitlis PM (1978) Acc Chem Res 11: 301; b) Wolzcanski PT, Bercaw JE (1980) ibid 13: 121; c) McLain SJ, Sancho J, Schrock RR (1979) J Am Chem Soc 101: 5451
16. a) Tolman CA (1977) Chem Rev 77: 313; b) Poe AJ (1988) Pure Appl Chem 60: 1209; c) Liu H-Y, Eriks K, Prock A, Giering WP (1990) Organometallics 9: 1758
17. a) Wilkinson G, Cotton FA (1959) Prog Inorg Chem 1: 1; b) Birmingham J (1964) Adv Organomet Chem 2: 365; c) Bublitz DE, Rinehart KL, Jr (1972) Org React 17: 1
18. Macomber DW, Hart WP, Rausch MD (1982) Adv Organomet Chem 21: 1
19. Coville NJ, Du Plooy KE, Pickl W Coord Chem Rev in press
20. Bruce MI, White AH (1990) Aust J Chem 43: 949
21. a) Poli R (1991) Chem Rev 91: 509; b) Baker RT in Inorganic reactions and methods, Zuckerman JJ, Hagen AP, (eds) (1991) VCH, Weinheim, Vol 12b: 35

22. See for example: Pedersen SF, Schrock RR, Churchill MR, Wasserman, HJ (1982) J Am Chem Soc 104: 6808; Schrock RR, Pedersen SF, Churchill MR, Ziller JW (1984) Organometallics 3: 1574
23. Riemschneider R (1963) Z Naturforsch, B 18: 641
24. a) Houben-Weyl, Methoden der Organischen Chemie, Vol 5/1c, p 664 Thieme, Stuttgart, 1970; b) Sitzmann H (1990) Z Naturforsch B 44: 1293
25. Venier CG, Casserly EW (1990) J Am Chem Soc 112: 2808
26. Maier G, Pfriem S, Schäfer U, Malsch K-D, Matusch R (1981) Chem Ber 114: 3965
27. a) Renaut P, Tainturier G, Gautheron B (1978) J Organomet Chem 148: 35; b) Freiesleben W (1963) Angew Chem 75: 576 Angew Chem Int Ed Engl (1963) 2: 396
28. Leigh T (1964) J Chem Soc 3294
29. Leonova EV, Kochetkova NS, Vainberg AM (1971) Zh Org Khim 7: 1912
30. Kaluski ZL, Gusev AI Kalinin AE, Struchkov YT (1972) Zh Strukt Khim 13: 950
31. Luke WD, Streitwieser A, Jr (1981) J Am Chem Soc 103: 3241
32. Bitterwolf TE, Ling AC (1977) J Organomet Chem 141: 355
33. El-Hinnawi MA, El-Khateeb MY, Jibril I, Abu-Orabi ST (1989) Synth React Inorg Met-Org Chem 19: 809
34. El-Hinnawi MA, Kobeissi MA (1989) Inorg Chim Acta 166, 99
35. Okuda J (1990) J Organomet Chem 385: C 39
36. Hofmann W, Buchner W, Werner H (1977) Angew Chem 89: 836; Angew Chem Int Ed Engl (1977) 16: 795
37. Werner H, Hofmann W (1981) Chem Ber 114: 2681
38. Nesmeyanov AN, Leonova EV, Kochetkova NS, Butyugin SM, Meisner IS (1971) Izv Akad Nauk SSSR 106
39. a) Urazowski IF, Ponomaryev VI, Ellert OG, Nifantev IE, Lemenovskii DA (1988) J Organomet Chem 356: 181; b) Urazowski IF, Ponomaryev VI, Nifantev IE, Leme-novskii DA (1989) ibid 368: 287
40. Bercaw JE, Marvich RH, Bell LG, Brintzinger HH (1972) J Am Chem Soc 94: 1219
41. Coutts RSP, Wailes PC, Martin RL (1973) J Organomet Chem 47: 375
42. a) Bürger H, Dämmgen U (1975) J Organomet Chem 101: 295; b) Dämmgen U, Bürger H (1975) ibid 101: 307
43. a) Sitzmann H (1990) Chem Ber 123: 2311; b) Sitzmann H (1988) J Organomet Chem 354: 203
44. Donovan BT, Hughes RP, Trujillo HA (1990) J Am Chem Soc 112: 7076
45. Gloaguen B, Astruc D (1990) J Am Chem Soc 112: 4607
46. Astruc D (1986) Acc Chem Res 19: 377
47. a) Mislow K (1986) Chimia 40: 395; b) Schuster II, Weissensteiner W, Mislow K (1986) J Am Chem Soc 108: 6661
48. a) Schumann H, Janiak C, Köhn RD, Loebel J, Dietrich A (1989) J Organomet Chem 365: 137; b) Schumann H, Görlitz FH, Dietrich A (1989) Chem Ber 122: 1423
49. Chambers JW, Baskar AJ, Bott SG, Atwood JL, Rausch MD (1986) Organometallics 5: 1635; Rausch MD, Tsai W, Chambers JW, Rogers RD, Alt HG (1989) Organometallics 8: 816
50. Stein, D, Sitzmann H (1991) J Organomet Chem 402: 249; 1991, 402: C 1
51. a) Hunter G, Weakley TJR, Mislow K, Wong MG (1986) J Chem Soc Dalton Trans 577; b) Downton PA, Mailvaganam B, Frampton CS, Sayer BG, McGlinchey MJ (1990) J Am Chem Soc 112: 27
52. Burk MJ, Arduengo AJ, Calabrese, JC, Harlow RL (1989) J Am Chem Soc 111: 8938
53. Pauson PL (1954) J Am Chem Soc 76: 2187
54. a) Carpenter, H (1898) Justus Liebigs Ann. Chem 302: 223; b) Ziegler K, Schnell B (1925) ibid 445: 266
55. Schumann H, Janiak C, Zuckerman JJ (1988) Chem Ber 121: 207
56. Cava MP, Narasimhan KJ (1969) J Org. Chem 34: 3641
57. a) Nakamura A, Hagihara N (1963) J Chem Soc Japan 84: 344; b) Feher F, Green M, Orpen GA (1986) J Chem Soc Chem Commun 291
58. a) Benn H, Wilke G, Henneberg D (1973) Angew Chem 85: 1052 Angew Chem Int (ed) Engl (1973) 12: 1001; b) Listeman ML, Schrock RR (1985) Organometallics 4: 74
59. Castellani MP, Wright JM, Geib SJ, Rheingold AL, Trogler WC (1986) Organometallics 5: 1116

60. Castellani MP, Geib SJ, Rheingold AL, Trogler WC (1987) Organometallics 6: 1703
61. a) McVey S, Pauson PL (1965) J Chem Soc 4312; b) Field LD, Hambley TW, Lindall CM, Masters AF (1989) Polyhedron 8: 2425
62. Evard PG, Piret P, Germain G, Van Meerssche, M (1971) Acta Cryst B 27: 661
63. a) Wakatsuki H, Yamazaki H (1985) Bull Chem Soc Japan 58: 2715; b) Bönnemann H (1985) Angew Chem 97: 264; Angew Chem Int (ed) (1985) 24: 248
64. Castellani MP, Geib SJ, Rheingold AL, Trogler WC (1987) Organometallics 6: 2524
65. Mao F, Philbin CE, Weakley TJR, Tyler DR (1990) Organometallics 9: 1510
66. Baukova TV, Slovokhotov YL, Struchkov YT (1981) J Organomet Chem 220: 125; (1981) 221: 375
67. Heeg, MJ, Janiak C, Zuckerman JJ (1984) J Am Chem Soc 106: 4259
68. Connelly NG, Manners I (1989) J Chem Soc Dalton Trans 283
69. Brégaint P, Hamon J-R, Lapinte C (1990) J Organomet Chem 328: C 25
70. Slocum DW, Duraj S, Matusz M, Cmarik JL, Simpson KM, Owen DA in Metal containing polymeric systems, Sheats JE, Carraher CE, Pittman CU (eds) (1985) Plenum, New York p 59
71. Brown KN, Field LD, Lay PA, Lindall CM, Masters AF (1990) J Chem Soc Chem Commun 408
72. Kläui W, Ramacher L (1986) Angew Chem 98: 107; Angew Chem Int (ed) Engl (1986) 25: 97
73. a) Connelly NG, Raven SJ (1986) J Chem Soc Dalton Trans 1613; b) Connelly NG, Raven SJ, Geiger WE (1987) ibid 467; c) Connelly NG, Geiger WE, Lane GA, Raven SJ, Rieger PH (1986) J Am Chem Soc 108: 6219
74. a) Lehmkuhl H, Naydowski C, Benn R, Rufinska A, Schroth G, Mynott R, Krüger C (1983) Chem Ber 116: 2447; b) Kölle U, Fuss B, Khouzami F, Gersdorf J (1985) J Organomet Chem 290: 77
75. Huhn M, Kläui W, Ramacher L, Herbst-Irmer R, Egert E (1990) J Organomet Chem 398: 339
76. Kölle U, Khouzami F (1980) Angew Chem 92: 658; Angew Chem Int (ed) Engl (1980) 19: 640
77. Schott A, Schott H, Wilke G, Brandt J, Hoberg H, Hoffmann EG (1973) Justus Liebigs Ann Chem 508
78. Hoberg H, Krause-Göing R, Krüger C, Sekutowski JC (1977) Angew Chem 89: 179; Angew Chem Int (ed) Engl (1977) 16: 183
79. a) Jack T, May CJ, Powell J (1977) J Am Chem Soc 99: 4707; b) Ban E, Cheng PT, Jack TR, Nyburg SC, Powell J (1973) J Chem Soc Chem Commun 368; c) Broadley K, Lane GA, Connelly NG, Geiger WE (1983) J Am Chem Soc 105: 2486; d) Broadley K, Connelly NG, Lane GA, Geiger WE (1986) J Chem Soc Dalton Trans 373
80. a) Powell J, Dowling NI (1983) Organometallics 2: 1742; b) Lane GA, Geiger WE, Connelly NG (1987) J Am Chem Soc 109: 402
81. Hübel W, Merenyi R (1964) J Organomet Chem 2: 213
82. Cox PA, Grebenik P, Perutz RN, Robinson MD, Grinter R, Stern DR (1983) Inorg Chem 22: 3614
83. a) Edelman MA, Lappert MF, Atwood JL, Zhang H (1987) Inorg Chim Acta 139: 185; b) Blake PC, Lappert MF, Taylor RG, Atwood JL, Zhang H (1987) ibid 139: 13
84. Jutzi P, Sauer R (1973) J Organomet Chem 50: C 29
85. Jutzi P (1986) Chem Rev 86: 983
86. Bassindale AR, Taylor PG (1989) in The chemistry of organic silicon compounds, Patai S, Rappoport Z, (eds) Wiley, New York p 893
87. Antinolo A, Bristow GS, Campbell GK, Duff AW, Hitchcock PB, Kamarudin RA, Lappert MF, Norton RJ, Sarjudeen N, Winterborn DJW, Atwood JL, Hunter WE, Zhang H (1989) Polyhedron 8: 1601
88. Antinolo A, de Ilarduya JM, Otero A, Royo P, Lanfredi AMM, Tiripicchio A (1988) J Chem Soc Dalton Trans 2685
89. Okuda J, Albach RW, Herdtweck E, Wagner FE Polyhedron in press
90. Rausch MD, Vogel M, Rosenberg H (1957) J Org Chem 22: 900

91. Tolstikov GA, Miftakhov MS, Monakov YB (1976) Zh Obshch Khim 46: 1778
92. Okuda J, Herdtweck E (1989) J Organomet Chem 373: 99
93. Okuda (1989) J Organomet Chem 367: C 1
94. Winter CH, Zhou X-X, Dobbs DA, Heeg MJ (1991) Organometallics 10: 210
95. Okuda J unpublished results
96. a) Jutzi P, Kuhn M (1979) J Organomet Chem 173: 221; b) Jutzi P, Seufert A (1979) J Organomet Chem 169: 373
97. Duff AW, Hitchcock PB, Lappert MF, Taylor RG, Segal JA (1985) J Organomet Chem 293: 271
98. Antinolo A, Lappert MF, Singh A, Winterborn DJW, Engelhardt LM, Raston CL, White AH, Carty AJ, Taylor NJ (1987) J Chem Soc Dalton Trans 1463
99. Okuda J (1988) J Organomet Chem 356: C 43
100. Lappert MF, Leung WP, Bartlett RA, Power PP (1989) Polyhedron 8: 1883
101. Antinolo A, Lappert MF, Lawless GA, Olivier H (1989) Polyhedron 8: 1882
102. Antinolo A, Lappert MF, Winterborn DJW (1984) J Organomet Chem 272: C 37
103. Miftakhov MS, Tolstikov GA (1976) Zh Obshch Khim 46: 930
104. Okuda J (1987) J Organomet Chem 333: C 41
105. Okuda J (1989) Chem Ber 122: 1259
106. Okuda J (1989) J Organomet Chem 375: C 13
107. Okuda J, Herdtweck E (1988) Chem Ber 121: 1899
108. Okuda J (1989) Chem Ber 122: 1075
109. Okuda J (1988) J Organomet Chem 353: C 1
110. Okuda J (1990) Chem Ber 123: 87
111. Okuda J, Herdtweck E (1991) Inorg Chem 30: 1516
112. Winter CH, Kampf JW, Zhou X-X (1990) Acta Cryst C 46: 1231
113. Gomez-Sal MP, Mena M, Royo P, Serrano R (1988) J Organomet Chem 358: 147
114. Winter CH, Dobbs DA, Zhou X-X (1991) J Organomet Chem 403: 145
115. Rosenblum M, Woodward RB (1958) J Am Chem Soc 80: 5443
116. a) Mann BE, in Comprehensive organometallic chemistry, Vol 3, Wilkinson G, Stone FGA, Abel EW (eds) (1982) Pergamon, Oxford, p 111; b) Holm CH, Ibers JA (1959) J Chem Phys 30: 885; c) Mulay LN, Attalla A (1963) J Am Chem Soc 85: 702; d) Makova MK, Leonova EV, Karimov YS, Kochetkova NS (1973) J Organomet Chem 55: 185; e) Gilson DFR, Gomez G (1982) J Organomet Chem 240: 41; f) Campbell AJ, Fyfe CA, Harold-Smith D, Jeffrey KR (1976) Mol Cryst Liq. Cryst 36: 1; g) Sorriso S, Cardaci G, Murgia SM (1972) J Organomet Chem 44: 181; h) Mann BE, Spencer CM, Taylor BF, Yavari P (1984) J Chem Soc Dalton Trans 2027; i) Adams H, Bailey NA, Mann BE, Taylor BF, White C, Yavari P (1987) ibid 1947; j) Eisenberg A, Shaver A, Tsutsui T (1980) J Am Chem Soc 1416
117. Mynott R, Lehmkuhl H, Kreuzer E-M, Joussen E (1990) Angew Chem 102: 314; Angew Chem Int (ed) Engl (1990) 29: 289
118. Haaland A (1974) Top Current Chem 53: 1
119. See for example: Zenneck U, Elschenbroich C, Möckel R (1981) J Organomet Chem 219: 177
120. See for example: Hu X, Duchowski J, Pomeroy, RK (1988) J Chem Soc Chem Commun 362
121. Du Plooy KE, Marais CF, Carlton L, Hunter R, Boeyens JCA, Coville NJ (1989) Inorg Chem 28: 3855
121. a) Erker G, Mühlenbernd T, Benn R, Rufinska A, Tsay Y-H, Krüger C (1985) Angew Chem 97: 336, Angew Chem Int (ed) Engl (1985) 24: 321; b) Erker G, Mühlenbernd T, Rufinska A, Benn R (1987) Chem Ber 120: 507; c) Erker G, Nolte R, Krüger C, Schlund R, Benn R, Grondey H, Mynott R (1989) J Organomet Chem 364: 119
122. Benn R, Grondey H, Erker G, Aul R, Nolte R (1990) Organometallics 9: 2493
123. Erker G, Nolte R, Tainturier G, Rheingold A (1989) Organometallics 8: 454
124. a) Ojima I, Clos N, Bastos C (1989) Tetrahedron 45: 6901; b) Bosnich B, Fryzuk MD (1981) Top Stereochem 12: 119
125. Bosnich B (1986) Asymmetric Catalysis (NATO ASI Series), Nijhoff, Dordrecht

126. Kagan H (1985) in Asymmetric synthesis, Morrison JD (ed) Academic, New York, Vol 5, p 1
127. Brunner H (1980) Adv Organomet Chem 18: 151
128. Thomson JB (1959) Tetrahedron Lett 6: 26
129. Schlögl K (1967) Topics Stereochem 1: 39; J Organomet Chem (1986) 300: 219
130. Ito Y, Sawamura M, Hayashi T (1986) J Am Chem Soc 108: 6405; Togni A, Pastor SD (1990) J Org Chem 55: 1649
131. Davies SG (1990) Aldrichim Acta 23: 31
132. Fernandez JM, Gladysz JA (1986) Inorg Chem 25: 2672
133. Faller JW, Lambert C (1985) Tetrahedron 41: 5755
134. Duthaler RO, Hafner A, Riediker M (1990) Pure Appl Chem 62: 631
135. Leblanc JC, Moise C, Bounthakna T (1974) Compt Rend Acad Sci Paris, Ser C 278 C, 973, 1975, 280 C 1431
136. See for example: Lindsay C, Cesarotti E, Adams H, Bailey NA, White C (1990) Organometallics 9: 2594
137. See for example: Nogradi M (1987) Stereoselective synthesis, VCH Weinheim
138. Cesarotti E, Ugo R, Vitiello R (1981) J Mol Cat 12: 63
139. Sato F, Iijima S, Sato M (1981) J Chem Soc Chem Commun 180
140. Cesarotti E, Kagan HB, Goddard R, Krüger C (1978) J Organomet Chem 162: 297
141. Couturier S, Tainturier G, Gautheron B (1981) J Organomet Chem 195: 291
142. Erker G, Nolte R, Tsay Y-H, Krüger C (1989) Angew Chem 101: 642; Angew Chem Int (ed) Engl (1990) 28: 628
143. Gleiter R, Paquette LA (1983) Acc Chem Res 16: 328
144. a) Paquette LA, Schirch PFT, Hathaway SJ, Hsu L-Y, Gallucci JC (1986) Organometallics 5: 490; b) Bhide V, Rinaldo P (1989) J Organomet Chem 376: 91
145. Galucci JC, Gautheron B, Gugelchuk M, Meunier P, Paquette LA (1987) Organometallics 6: 15
146. Paquette LA, Moriarty KJ, Meunier P, Gautheron B, Crocq V (1988) Organometallics 7: 1873
147. Halterman RL, Vollhardt KPC (1986) Tetrahedron Lett 1461
148. a) Paquette LA, McKinney JA, McLaughlin ML, Rheingold AL (1986) Tetrahedron Lett 27: 5599; b) McLaughlin ML, McKinney, JA Paquette LA (1986) ibid 27: 5595
149. Halterman RL, Vollhardt KPC (1988) Organometallics 7: 883
150. a) Paquette LA, Moriarty KJ, McKinney JA, Rogers RD (1989) Organometallics 8: 1707; b) Paquette LA, Gugelchuk M, McLaughlin ML (1987) J Org Chem 52: 4732
151. a) Moriarty KJ, Rogers RD, Paquette LA (1989) Organometallics 8: 1512; b) Sivik MR, Rogers RD, Paquette LA (1990) J Organomet Chem 397: 177
152. Paquette LA, Moriarty KJ, Rogers RD (1989) Organometallics 8: 1506
153. Whitesell JK (1989) Chem Rev 89: 1581
154. a) Halterman RL, Vollhardt KPC, Welker ME, Blaser D, Boese R (1987) J Am Chem Soc 109: 8105; b) Chen ZL, Halterman RL (1990) Synlett 2: 103
155. Coletti SL, Halterman RL (1989) Tetrahedron Lett 30: 3513
156. Erker G, van der Zeijden AAH (1990) Angew Chem 102: 543; Angew Chem Int (ed) Engl (1990) 29: 512; Erker G (1990) J Organomet Chem 400: 185
157. Kaminsky W, Külper K, Brintzinger HH, Wild FRWP (1985) Angew Chem 97: 507; Angew Chem Int (ed) Engl (1985) 24: 507
158. See for example: Spaleck W, Antberg M, Dolle V, Klein R, Rohrmann J, Winter A (1990) New J Chem 14: 499
159. Schnutenhaus H, Brintzinger HH (1979) Angew Chem 91: 837; Angew Chem Int (ed) Engl (1979) 18:777
160. Wiesenfeldt H, Reinmuth A, Barsties E, Evertz K, Brintzinger HH (1989) J Organomet Chem 369: 359
161. Mise T, Miya S, Yamazaki H (1989) Chem Lett 1853
162. Röll W, Brintzinger HH, Rieger B, Zolk R (1990) Angew Chem 102: 339; Angew Chem Int (ed) Engl (1990) 29: 279
163. Collins S, Hong Y, Taylor NJ (1990) Organometallics 9: 2695
164. Abel EW, Long NJ, Orell KG, Osborne AG, Sik V (1991) J Organomet Chem 403: 195

Author Index Volumes 151–160

Author Index Vols. 26–50 see Vol. 50
Author Index Vols. 50–100 see Vol. 100
Author Index Vols. 101–150 see Vol. 150

The volume numbers are printed in italics